The Water–Energy–Food Nexus in the Middle East and North Africa

This book discusses key issues concerning water, energy and food in the Middle East and North Africa (MENA) region. It provides an interdisciplinary account of current developments in one of the most water-scarce and conflict-torn regions in the world. Key analysts on MENA water, agriculture and energy affairs have been drawn together to compile one of the first edited volumes dedicated to the crucial role of water, energy and food security in the 21st century MENA region. It will be of interest to decision makers, analysts and students of the future of the Middle East from a broad range of disciplines including the physical and social sciences. This book was previously published as a special issue of the *International Journal of Water Resources Development*.

Martin Keulertz is a researcher at King's College London and a visiting research fellow at Texas A&M University, USA. His research centres around the politics of water and food management in the MENA region.

Eckart Woertz is a senior research fellow at CIDOB, the Barcelona Centre for International Affairs, Spain, and scientific advisor to the Kuwait Chair at Sciences Po in Paris. His previous work has been on Gulf financial markets, food security and oil politics.

Routledge Special Issues on Water Policy and Governance

https://www.routledge.com/series/WATER

Edited by
Cecilia Tortajada (IJWRD), *Third World Centre for Water Management, Mexico*
James Nickum (WI), *International Water Resources Association, France*

Most of the world's water problems, and their solutions, are directly related to policies and governance, both specific to water and in general. Two of the world's leading journals in this area, the *International Journal of Water Resources Development* and *Water International* (the official journal of the International Water Resources Association), contribute to this special issues series, aimed at disseminating new knowledge on the policy and governance of water resources to a very broad and diverse readership all over the world. The series should be of direct interest to all policymakers, professionals and lay readers concerned with obtaining the latest perspectives on addressing the world's many water issues.

Water Pricing and Public-Private Partnership
Edited by Asit K. Biswas and Cecilia Tortajada

Water and Disasters
Edited by Chennat Gopalakrishnan and Norio Okada

Water as a Human Right for the Middle East and North Africa
Edited by Asit K. Biswas, Eglal Rached and Cecilia Tortajada

Integrated Water Resources Management in Latin America
Edited by Asit K. Biswas, Benedito P. F. Braga, Cecilia Tortajada and Marco Palermo

Water Resources Management in the People's Republic of China
Edited by Xuetao Sun, Robert Speed and Dajun Shen

Improving Water Policy and Governance
Edited by Cecilia Tortajada and Asit K. Biswas

Water Quality Management
Present Situations, Challenges and Future Perspectives
Edited by Asit K. Biswas, Cecilia Tortajada and Rafael Izquerdo

Water, Food and Poverty in River Basins
Defining the Limits
Edited by Myles J. Fisher and Simon E. Cook

The Water–Energy–Food Nexus in the Middle East and North Africa

Edited by
Martin Keulertz and Eckart Woertz

Taylor & Francis Group

LONDON AND NEW YORK

First published 2017 by Routledge

2 Park Square, Milton Park, Abingdon, Oxfordshire OX14 4RN
711 Third Avenue, New York, NY 10017

Routledge is an imprint of the Taylor & Francis Group, an informa business

First issued in paperback 2018

British Library Cataloguing in Publication Data
A catalogue record for this book is available from the British Library

ISBN 13: 978-1-138-67422-6 (hbk)
ISBN 13: 978-0-367-02844-2 (pbk)

Typeset in Times
by diacriTech, Chennai

Publisher's Note
The publisher accepts responsibility for any inconsistencies that may have arisen
during the conversion of this book from journal articles to book chapters, namely
the possible inclusion of journal terminology.

Disclaimer
Every effort has been made to contact copyright holders for their permission to
reprint material in this book. The publishers would be grateful to hear from any
copyright holder who is not here acknowledged and will undertake to rectify any
errors or omissions in future editions of this book.

Contents

CONTENTS

Citation Information

The chapters in this book were originally published in the *International Journal of Water Resources Development*, volume 31, issue 3 (September 2015). When citing this material, please use the original page numbering for each article, as follows:

Chapter 1

The water–energy–food nexus: an introduction to nexus concepts and some conceptual and operational problems
Tony Allan, Martin Keulertz and Eckart Woertz
International Journal of Water Resources Development, volume 31, issue 3 (September 2015) pp. 301–311

Chapter 2

Financial challenges of the nexus: pathways for investment in water, energy and agriculture in the Arab world
Martin Keulertz and Eckart Woertz
International Journal of Water Resources Development, volume 31, issue 3 (September 2015) pp. 312–325

Chapter 3

Food-water security and virtual water trade in the Middle East and North Africa
Marta Antonelli and Stefania Tamea
International Journal of Water Resources Development, volume 31, issue 3 (September 2015) pp. 326–342

Chapter 4

Rapid assessment of the water–energy–food–climate nexus in six selected basins of North Africa and West Asia undergoing transitions and scarcity threats
Caroline King and Hadi Jaafar
International Journal of Water Resources Development, volume 31, issue 3 (September 2015) pp. 343–359

Chapter 5

The nexus as a political commodity: agricultural development, water policy and elite rivalry in Egypt
Harry Verhoeven
International Journal of Water Resources Development, volume 31, issue 3 (September 2015) pp. 360–374

CITATION INFORMATION

Chapter 6

Nexus meets crisis: a review of conflict, natural resources and the humanitarian response in Darfur with reference to the water–energy–food nexus
Brendan Bromwich
International Journal of Water Resources Development, volume 31, issue 3 (September 2015) pp. 375–392

Chapter 7

To what end? Drip irrigation and the water–energy–food nexus in Morocco
Guy Jobbins, Jack Kalpakian, Abdelouahid Chriyaa, Ahmed Legrouri and El Houssine El Mzouri
International Journal of Water Resources Development, volume 31, issue 3 (September 2015) pp. 393–406

Chapter 8

Virtual-water content of agricultural production and food trade balance of Tunisia
Jamel Chahed, Mustapha Besbes and Abdelkader Hamdane
International Journal of Water Resources Development, volume 31, issue 3 (September 2015) pp. 407–421

Chapter 9

Energy cost of irrigation policy in Morocco: a social accounting matrix assessment
Mohammed Rachid Doukkali and Caroline Lejars
International Journal of Water Resources Development, volume 31, issue 3 (September 2015) pp. 422–435

Chapter 10

Impact of the Syrian conflict on irrigated agriculture in the Orontes Basin
Hadi H. Jaafar, Rami Zurayk, Caroline King, Farah Ahmad and Rami Al-Outa
International Journal of Water Resources Development, volume 31, issue 3 (September 2015) pp. 436–449

Chapter 11

Climate change and food-water supply from Africa's drylands: local impacts and teleconnections through global commodity flows
Mark Mulligan
International Journal of Water Resources Development, volume 31, issue 3 (September 2015) pp. 450–460

Chapter 12

Towards a water–energy–food nexus policy: realizing the blue and green virtual water of agriculture in Jordan
Samer Talozi, Yasmeen Al Sakaji and Amelia Altz-Stamm
International Journal of Water Resources Development, volume 31, issue 3 (September 2015) pp. 461–482

For any permission-related enquiries please visit:
http://www.tandfonline.com/page/help/permissions

Notes on Contributors

Farah Ahmad is a research assistant in the Department of Agriculture and Department of Landscape and Ecosystem Management, Faculty of Agricultural and Food Sciences, American University of Beirut, Lebanon.

Tony Allan is an Emeritus Professor in the Department of Geography at King's College London, and a Professor in the Department of Financial and Management Studies at SOAS, UK.

Rami Al-Outa is a landscape architect at Imad Gemayel Architects, Beirut, Lebanon.

Yasmeen Al Sakaji is a researcher in Bioenvironmental and Irrigation Engineering, Jordan University of Science and Technology, Irbid, Jordan.

Amelia Altz-Stamm is a researcher in the Lyndon B. Johnson School of Public Affairs, University of Texas at Austin, USA.

Marta Antonelli is a graduate teaching assistant at Università degli Studi Roma Tre, Italy.

Mustapha Besbes is an Emeritus Professor at the Université Tunis El Manar, Ecole Nationale d'Ingénieurs de Tunis, Tunisia.

Brendan Bromwich is a researcher in the Department of Geography, King's College London, UK.

Jamel Chahed is a Professor of Civil Engineering at the Université Tunis El Manar, Ecole Nationale d'Ingénieurs de Tunis, Tunisia.

Abdelouahid Chriyaa is a research director at the Institut National de la Recherche Agronomique, Settat, Morocco.

Mohammed Rachid Doukkali is a Professor in the Social Science Department, Institut National Agronomique Hassan II (IAV Hassan II) and a senior non-resident fellow of the OCP Policy Center, Morocco.

El Houssine El Mzouri is the Director of the Institut National de la Recherche Agronomique, Settat, Morocco.

Abdelkader Hamdane is a researcher at the Universite´ de Carthage, Institut National Agronomique de Tunis, Tunisia.

Hadi Jaafar is an Assistant Professor in the Department of Agriculture, Faculty of Agricultural and Food Sciences, American University of Beirut, Lebanon.

NOTES ON CONTRIBUTORS

Guy Jobbins is a research fellow at the Overseas Development Institute, London, UK.

Jack Kalpakian is an Associate Professor in the School of Humanities and Social Sciences, Al Akhawayn University in Ifrane, Morocco.

Martin Keulertz is a researcher at King's College London and a visiting research fellow at Texas A&M University, USA. His research centres around the politics of water and food management in the MENA region.

Caroline King is a graduate student in the Centre for the Environment, School of Geography and the Environment, Oxford University, UK.

Caroline Lejars is a Visiting Associate Professor in the Social Science Department, Institut National Agronomique Hassan II (IAV Hassan II), Morocco.

Ahmed Legrouri is a Professor of Chemistry in the School of Science and Engineering, Al Akhawayn University in Ifrane, Morocco.

Mark Mulligan is a Reader in Geography at the Department of Geography, King's College London, UK.

Samer Talozi is an Assistant Professor in the Civil Engineering Department, Jordan University of Science and Technology, Irbid, Jordan.

Stefania Tamea is a post-doctoral researcher in the Department of Environment, Land and Infrastructure Engineering, Politecnico di Torino, Turin, Italy.

Harry Verhoeven is a post-doctoral fellow in the Department of Politics & International Relations, University of Oxford, UK, and an Assistant Professor in the School of Foreign Service (Qatar), Georgetown University, Washington, DC, USA.

Eckart Woertz is a senior research fellow at CIDOB, the Barcelona Centre for International Affairs, Spain, and scientific advisor to the Kuwait Chair at Sciences Po in Paris. His previous work has been on Gulf financial markets, food security and oil politics.

Rami Zurayk is a Professor in the Department of Landscape and Ecosystem Management, Faculty of Agricultural and Food Sciences, American University of Beirut, Lebanon.

The water–energy–food nexus: an introduction to nexus concepts and some conceptual and operational problems

Tony Allan[a], Martin Keulertz[b] and Eckart Woertz[c]

[a]King's College London and SOAS, UK; [b]Purdue University, West Lafayette, Indiana, USA;
[c]Barcelona Centre for International Affairs (CIDOB), Spain

This introduction sets the scene for the special issue compiled by Martin Keulertz, Eckart Woertz and Tony Allan.

The conceptual deficit

The purpose of this special issue is to provide an introduction to the concept of the water–energy–food nexus and the challenges faced by those wanting to use the idea in global drylands. This introductory article will take a holistic, global perspective. Globally, water resources are limited and effectiveness of use is determined by technology and management. But water use trends cannot easily be altered as water demand is determined by demography and consumption behaviour. Water consumption is likely to increase by 55% worldwide between 2000 and 2050. Some industries will increase their water use by 400% (manufacturing) and by 140% (electricity) (Organisation for Economic Cooperation and Development [OECD], 2012). Using a nexus approach to steward water resources sustainably in energy supply chains and food supply chains is seen as a promising approach.

But problems have been encountered in its initial deployment. One of the most important tasks facing society, whether in drylands or in other climate zones, is to find ways to value water to operationalize the nexus. The ways that natural resources such as water and energy have been developed and consumed by society have *not* been shaped by awareness of their scarcity or their value. It will be emphasized that there are major asymmetries in the consumption of water and energy. For example, farmers manage 92% of the water resources consumed in food supply chains (Hoogeveen, 2015). How they manage and mismanage water impacts the atmosphere and water and other ecosystems. In these circumstances society needs farmers to be or to become good stewards of natural ecosystems. But the private-sector food supply chains that connect consumers with those who produce food are dysfunctional in this regard. Those who manage current food supply chains are not ecosystem-aware and are not precautionary with respect to the sustainability of the natural ecosystems on which the food security of society depends.

There are four reasons for this. First, food is very emotional and food prices are very highly politicized. Food supply chains are impacted by subsidies that are of vital significance to poor elements of society. Secondly, there are serious asymmetric power relations in food supply chains; farmers are generally economically weak despite the strength of farm lobbies in OECD countries. Thirdly, those with power in private-sector food supply chains – the corporations – handle a very small proportion of the embedded natural resources. They have potential contractual leverage over farmers who do manage vast volumes of water, but they have as yet little incentive to engage outside the fence of their warehouses, silos, factories and wineries. Fourthly, market signals and the reporting and accounting systems that track them are dangerously partial and blind to the values of natural resource inputs, especially of water; and they do not capture the costs of mismanaging them.

By the mid-twentieth century major, often very long, international private-sector supply chains had been established. Very important long international supply chains – in which water and energy are embedded – have existed for over a century. They have supplemented in major ways the private-sector sub-national food and energy supply chains that have secured food and energy for the world's increasingly urbanized populations.

The food and energy supply chains are both market systems, but they have taken very different approaches to valuing water, food and energy. Food and energy have always been of high strategic significance to society and to the governments that are responsible for their security. The ways that food and energy are priced are the result of existential processes in the political economies of food and energy. Both food and energy are also extremely emotional at all levels of social organization. They are also deeply embedded in the social contract between society and those who govern. As a consequence, the tools available to states in intervening in these political economies – taxes and subsidies – feature very prominently in food and energy policies. Once in place they are even more difficult to remove than they were to install. Those in power have judged that they can best stay in power by ensuring that their poorest citizens enjoy access to cheap food and stable energy prices. As a consequence, food supply chains are associated with a myriad of direct production subsidies, for example those of the EU Common Agricultural Policy regime and of the US Farm Bill. In many low-income economies the subsidies are indirect, through the provision of subsidized diesel or electricity to pump water for irrigation.

At this point in history, at the beginning of the twenty-first century, the outcome of the past 200 years of evolution of the political economies of food (Paarlberg, 2013) and of energy (Pascual & Elkind, 2009) is that global food prices are determined by massive policy interventions by OECD governments and those in emerging and developing economies. Water resources have been dangerously impacted by these interventions because it has been assumed that there were no ecosystem costs associated with over-using, depleting and polluting unvalued or under-valued environmental resources and services.

In the case of energy, prices have been even more distorted. Until about 1980 it was assumed in the energy sector that consumption and misuse – in this case of fossil fuels – were environmentally costless. But consumers of dirty coal, and especially more recently of dirty oil and gas, have been prepared to pay prices that have been loaded with taxes, while at the same time their economies enjoyed the fall in the constant price of oil that characterized the period from 1900 to 1972. In the early 1980s, the hegemons of the global energy regime were able to reintroduce the downward trend in oil prices of the pre-1972 era. At the millennium, global oil prices tightened and spiked in 2008 and 2011. It remains to be seen whether this will be a sustained reversal of the trend of declining prices in real terms that prevailed over much of the 20th century. Doing the wrong thing extremely well is a common feature of capitalism. In such processes society can enjoy some short-term benefits, but labour, the atmosphere and all natural ecosystems have paid a very high price.

Global food prices have been determined by farm politics, food politics and food policies in rich OECD economies. Energy prices have been determined by the alliance of big-oil corporations and major OECD governments that was able to manipulate pivotal producers. Food producers in emerging and developing economies have been very negatively impacted by the price regimes determined by the global political economy of food production and trade. In both supply chains the interests of the OECD economies were very influential. But the two supply chains are in many respects very different, especially at the beginning. In the oil and gas supply chains there is no equivalent to the half-billion or so farmers, mainly on small commercial farms and on subsistence farms. In the food supply chains there is no equivalent to the national oil companies in the major oil-exporting economies or the exploration and marketing companies of the OECD and emerging economies.

It will be a recurring theme in this analysis that the scarcity values of water and energy embedded in food and manufactured commodities are not reflected in the prices paid by consumers for the goods they purchase in private-sector markets. Nor are the use values of water and energy captured. Because the exchange values along the supply chains have been very severely distorted by subsidies and taxes, the costs of degrading water and other ecosystem services have been invisible and until recently ignored.

The emergence of the sub-nexi and the grand nexus

In 2008, the World Economic Forum – to the activities of which Tony Allan, one of the authors of this article, contributed as a member of its Water Advisory Council – identified water, food, energy and climate change as a nexus where risks for society, the economy and the environment were very serious. The forum concluded that the risks were being multiplied by ignoring the contradictions of current rates of water and energy use in hot spots and more generally.

The World Economic Forum reviewed the water–food–trade sub-nexus and the energy–climate change sub-nexus and an attempt was made to address the integration of these two sub-nexi into a grand nexus. The WEF publication on the nexus (Waughray, 2011) was rich in scope, but the editorial team and its contributors – of which Tony Allan was one – failed to provide an accessible framework that identified the important key assumptions, key issues, and key players in the key private-sector supply chains. Nor did the publication identify incentives that would bring together the two very powerful operational supply chains responsible for the way society was using and abusing its natural resources of water and energy.

In another potentially important initiative in the 2010 and 2012 period, two government ministries of the Federal Republic of Germany convened a suite of international nexus meetings attended by numerous scientists, as well as by government, private-sector and NGO professionals. Tony Allan participated in a minor way. The courteous convenors allowed him to contribute a critique of the frenetic engagement on the grand nexus at that time. He argued again that a profound and useful conceptualization of the grand nexus was lacking. He suggested that the absence of an overarching theoretical frame was making it impossible for those engaging to communicate effectively. He pointed out again that the water–food–trade sub-nexus had been effectively conceptualized, but the energy–climate change sub-nexus and a theorization of the water and energy sub-nexi into a grand nexus had not been achieved. He also argued that the operations of private-sector food supply chains and energy supply chains had determined the priorities as well as the blind spots of the business-as-usual operations of the sub-nexi and the grand nexus. It was in these private-sector practices that water and energy were being managed and mismanaged. It was also noted that these major supply chain players would also be the key agents that could most effectively analyze, and subsequently engage, to address the current contradictions that were becoming evident as a consequence of the attempts to develop a grand nexus approach.

How the sub-nexi and the grand nexus fit with operational political economies of water and energy

It is important to take some initial steps in establishing a conceptual frame that embraces the two sub-nexi – that is, the *water–food–trade sub-nexus* and *the energy–climate change* sub-nexus. These two sub-nexi need to be integrated into an as yet poorly understood *grand nexus,* where both interact. It will be concluded that the grand nexus is one way of characterizing parts – albeit important parts – of the overarching political economy where production, trading, processing and marketing activities deliver goods and services to consumers. Awareness of this hierarchy of engagement is essential in any consideration of the grand nexus of water, food and energy. Deconstructing dynamic global and regional market operations is likely to be inconclusive at best, and at worst it will be very misleading. Private-sector firms are constantly merging and demerging both within and between numerous supply chains, of which the ones we are analyzing here are only two.

Those who operate the private-sector supply chains of food and energy are best acquainted with the challenges in the sub-nexus in which they operate (British Petroleum, 2013). They will be the agents that know the most about their respective sub-nexus operations and constraints. They will also be the agents who could potentially identify and adapt to the risks of ignoring the mutualities and contradictions integral to grand nexus operations. Farmers, big-ag corporations and other major corporations in food, trade, supermarkets, oil and gas, and vehicles and transport will all be key players. Subsidies, taxes and the poorly informed choices of consumers in both sub-nexi supply chains will also play significant roles in steering decision making in the as yet uncoordinated sub-nexus supply chains.

The grand nexus of water, energy and food is dominated by market mechanisms and supply value chains that are not yet equipped to expose the environmental and social risks associated with the otherwise rather effective market systems that produce and provide foods and services. Unfortunately, these market systems fatally lack reporting and accounting rules that could capture the costs of natural ecosystems – atmosphere, water and biodiversity. Such rules would signal to those who manage vulnerable and sometimes very scarce water and energy resources the consequences of not understanding natural resource scarcities and values. In depleting and potentially damaging natural resource inputs in food and energy supply chains, markets have risked the planet, economies and society as well as the sustainability of their own enterprises. It has proved to be easier to get their attention on the risk to the profits of business as usual, and in the past decade there is much evidence of corporate awareness of their vulnerabilities to both local water and global energy scarcities.

The approach suggested here requires a different point of departure from that which has resulted from the assumptions of science and of water and energy professionals. Scientists and economists assume that there is some rational, knowledge-based, potentially optimizable way of allocating and managing water and energy. In contrast, it is assumed here that farmers, manufacturers and other market players have operated the water–food–trade sub-nexus for at least two millennia. It is further assumed that long-established private-sector supply chains have determined the energy–climate change sub-nexus. Separate market systems using very different proportions of embedded water and energy deliver food commodities and energy to consumers (Mekonnen and Hoekstra, 2011). As yet these separate supply chains have not engaged effectively over the potential benefits of adopting a wider understanding of competition and of mutuality with respect to profit. Profit is their statutory concern and their duty to deliver. Responsible approaches to people and the planet are also needed, but they can only protect them to the extent that existing inadequate regulatory regimes require. As yet there are very few reporting rules and no accounting rules by which to steer.

A very useful preliminary to identifying and mapping the grand nexus and two sub-nexi is the question: What are the proportions of water and energy used in different sectors of the world's economies? There are major asymmetries in the use of water by economic sectors in the water–food–trade and energy–climate change sub-nexi. Huge volumes of water are utilized in food supply chains. Consumers and even water professionals are unaware that about 92% of *consumptive* water use is devoted to food production. Only about 10% of water is consumed by society in its industries and for domestic services. In a few economies, for example in Singapore and Israel, much of this water is in practice recycled, and it could potentially be recycled in most economies. All the water used for the production of non-food commodities and services is blue water.

Blue water is available in rivers, lakes, reservoirs and aquifers. Blue water can be diverted, pumped, conveyed and polluted and is the main concern of hydraulic engineers and of water resources and water quality scientists. Blue water can also be valued. It can be priced, and it is usually priced, but unfortunately very frequently mispriced, which makes it the focus of economists (Abu-Zeid, 2001; Biswas & Tortajada, 2010).

Green water comprises most of the water used to produce food and fibre in private-sector food supply chains. About 70% of the water used in food supply chains is green water. Green water is the water held in the soil profile after rainfall. Natural vegetation and crops draw on this water and transpire it to the atmosphere. All this water is lost to local users. It is recycled to the atmosphere in the global hydrological system, but it cannot be recycled locally. Green water is an essential resource in food production, and it also provides important ecosystem services (Falkenmark & Rockström, 2004; Hoff, 2011). Blue water in agriculture is different. It is used for irrigation and competes with other water users in industry, in other services, in the provision of recreational amenities and in the provision of domestic water services. The very vulnerable ecosystem services of blue water are increasingly highly prized, but their value is not yet captured.

As with green water, farmers are the main blue water users. Again, their role as managers and stewards of water, in this case of blue water, is not recognized by engineers, water scientists, economists, or society more generally. Food consumers are especially under-informed.

The market systems that connect consumers with blue and green water are dysfunctional. They are very complex, and in many ways very effective. They deliver vast volumes of food through long international and short local supply chains. But they are rough and ready in the sense that they recognize many but not all of the inputs in very selective and very partial accounting systems. Contracted prices connect the agents in the food supply chain markets but they do not incentivize the stewardship of water ecosystems or other ecosystem services.

Water is not effectively priced or not priced at all as an input, nor are the costs of polluting and misusing it captured. Green water is everywhere regarded as a free good. Blue water is usually regarded as a free good in the production of food and fibre, and is never properly priced. In the past two decades, however, for irrigated farming, some semi-arid OECD economies – in Israel and Australia for example – have installed tariff and market arrangements that have begun to link the costs of blue water in both utilizing and stewarding it (Gilmont, 2014).

Green and blue water differ greatly in their vulnerability to depletion and over-use. Green water can be said to look after itself. Farmers cannot use last year's soil-profile water. Nor can they use the moisture of the year before. They certainly cannot borrow moisture from next year's rains, nor from those of the following year. In contrast, blue water can be managed to use water from earlier years, stored in engineered reservoirs. Farmers can also pump blue water from groundwater aquifers that may not be replenished in the following seasons. Wherever farmers have irrigated they have run out of blue water, and the main reason we have debates on water security is that water used for irrigation is already competing very seriously in hot spots world-wide with other demands for blue water. Water used in other sectors – such as industry, mining and services – can command prices orders of magnitude higher than farmers can pay in food supply systems. In the food supply chain consumers, legislators and markets conspire to provide underpriced cheap food where the cost of water cannot be considered. There is also increasing competition for water in many hot spots worldwide, where the need to restore the ecosystem services of blue water has become vital.

There is only one type of non-renewable water resource: fossil water. These water resources accumulated in past geological eras. They are limited. Unfortunately, they are mostly short-lived economic phenomena, with an unknown life span (Famigletti, 2014). In contrast, there are at least six types of non-renewable energy: peat, brown coal, oil, tar sands, natural gas, shale gas and nuclear.

The difference is just as evident in the domain of renewable natural resources. There are two types of renewable water resources: green and blue. Again, renewable energy has at least six sources: wood, hydro, wind, solar, tide and bioenergy. Animal power used to be a very significant source of renewable energy. There are two other types of water that have as yet a small role in volumetric terms. Water can be manufactured with desalination technologies, and it can be recycled after use in industrial and domestic services. These types of water will play an increasingly important role in sustaining cities and their industries and services.

Private-sector supply chains: a frame that provides important insights

Having established the types of water and energy available and the volumes used in the two supply chains, the next challenge is to highlight the importance of recognizing that water and energy are mobilized and used in private-sector supply chains that have inadequate reporting and accounting rules.

Figure 1 shows a widely recognized way of understanding the structure of a neoliberal society and provides an easy-to-communicate approach to analyzing advanced political economies. Other political economies – emerging and developing – are also important, but this chapter is at an early stage of an analysis that is very much work in progress.

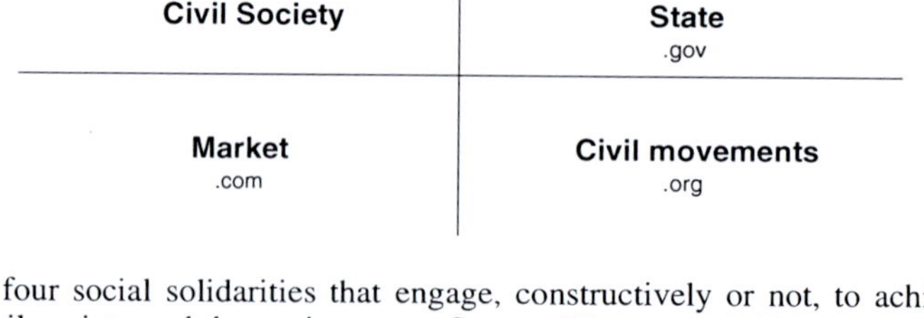

Civil Society	**State** .gov
Market .com	**Civil movements** .org

Figure 1. The four social solidarities that engage, constructively or not, to achieve sustainable outcomes for civil society and the environment. Source: Thompson (1990).

Figure 1 shows that there are three what social theorists (unhelpfully) call 'social solidarities', which do things to and for a fourth solidarity, namely civil society. Civil society is all of us at breakfast time. The three solidarities that do things such as providing food to civil society are the *public* (state), the *private* (market), and collective civil movements such as unions and NGOs.

The structure set out in Figure 1 is useful for mapping one of the sub-nexi considered in the introduction. Figure 2 shows how the water–food–trade private-sector market supply chain or sub-nexus can been mapped onto the 'four ways of life' structure shown in Figure 1.

The most important thing to note is that the water–food–trade sub-nexus maps very well on to the left half of the 'four ways of life' structure. First, all the agents involved in the water–food–trade supply chains can be accommodated within civil society, where food demands and wasteful food choices and wasteful behaviour are the norm; secondly, the private-sector market systems operated by commercial farmers, corporate commodity traders and processors, as well as by local shops and corporate supermarkets, can be mapped onto the bottom left of the diagram. They engage with each other via contracts, which ignore or mis-price the value of natural resource inputs. Subsistence farmers do not engage in these arrangements.

One of the asymmetries identified above – the massive volumes of water devoted to food production – can also be illustrated. It follows that 92% of the *consumptive* use of water by society takes place in private-sector food supply chains (Hoff, 2011). It is important to highlight the importance of consumptive water use. While 70% of global water is allocated to agriculture, 20% is allocated to industry and 10% to domestic users for water and sanitation practices. Consumptive use is defined as gross withdrawals minus return flow (Table 1). However, while industrial and domestic water can be reused through recycling methods, the actual consumptive use is 5% for industry and 3% for households

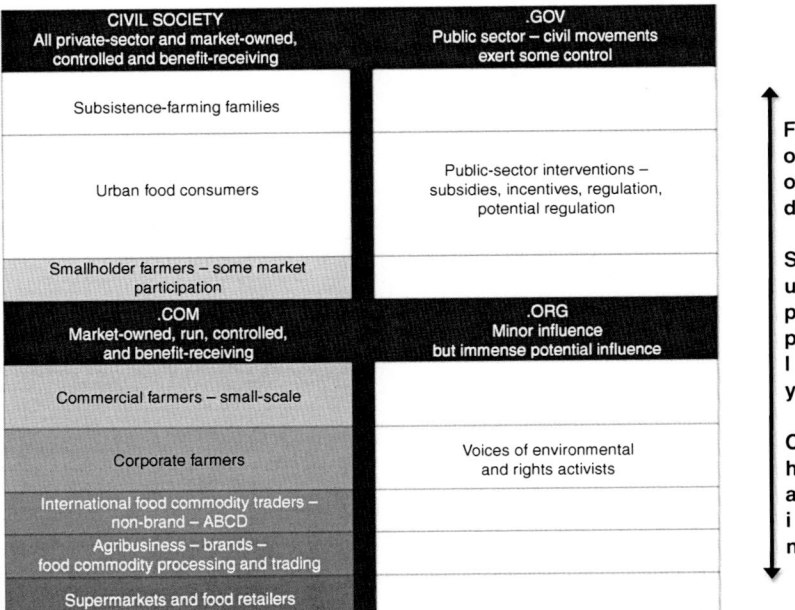

Figure 2.   The four ways of life, with the private-sector food supply sub-nexus mapped onto it. Note: ABCD, Archers Daniels Midlands, Bunge, Cargill and Louis Dreyfus. Source: Thompson (1990).

7

Table 1. Consumptive use of agriculture, industry and drinking water. Source: Hoogeveen (2015).

	Gross withdrawals (A)	Return flow (B)	Consumptive use (A × B)	Consumptive use (%)
Agriculture	70%	50%	0.35	92%
Industry	20%	10%	0.02	5%
Drinking water	10%	10%	0.01	3%
Total			0.38	

(Hoogeveen, Faurès, Peiser, Burke, & van de Giesen, 2015). The water in food cannot be reused, hence the 92% consumed by society. The consumptive use by agriculture includes green and blue water (rainfed farming and irrigation). This water is managed and mismanaged by farmers, who select water management practices. This supply chain operates as a market, but in the absence of the necessary reporting and accounting rules it is blind to water value and externalities.

Mainly public-sector service providers handle the other 10% of water, which is all blue water. Local authorities and other municipal bodies provide domestic and industrial water services worldwide (Allan, 2011). They increasingly have statutory responsibility for water and other ecosystem services under the umbrellas of national environmental protection regulatory authorities. The privatization of non-food water and sewage services has a recent chequered history, which is not the concern of this analysis. Privatization of these services is gaining ground, but very gradually.

The conceptualization presented in this section can be summarized as follows. There are very powerful, long-established private-sector supply chains. The complex and separate logistics of the respective supply chains are managed by extremely well-informed and increasingly well-equipped professionals from farm to supermarket and from oil well to consumer. Collectively, professional farmers, commodity traders and processors and supermarkets know more about their respective sub-nexus than any scientist or economist or government. Other private-sector professionals operate the energy sub-nexus markets.

The water–food–trade sub-nexus and the energy–climate change sub-nexus are separate and very different in character. They both ultimately serve seven billion global consumers, but this is one of the very few things they have in common. There is no equivalent in the energy supply chain nexus to the half-billion farmers at the beginning of the water–food–trade supply chain nexus. Nor is there an equivalent in the food and water supply chain to the spot-and-forward pricing systems that operate in global energy markets. There is pricing for food, but not for water in food. Some unsteady steps are being taken in the Murray-Darling Basin in Australia to set up equivalent systems, but the volumes of water involved are as yet trivial (Gilmont, 2014).

The powerful supply chain corporations operate in separate markets in the separate nexi. If oil companies owned irrigated farms, or if farmers were intimately involved in mining, there would be an automatic awareness of the unsustainable competitive utilization of water and of its different values in alternative uses. Their accountants would be aware and alert to profit-impacting contradictions. Awareness of the role of what we have identified as the *grand nexus* would be operationalized without hesitation. In practice, the operational and commercial separation of the two supply chains means that there is no incentive to know about the dangers of using unvalued and unpriced water resources.

It is suggested that the tendency of adopting a hot-spots approach will prevail as hot spots become evident. They have, for example, become evident in the water resource use contradictions associated with developing shale gas and tar/oil sands as well as in producing ethanol from corn. This hot-spots approach will continue until the private-sector players who operate rather than conceptualize the complex food and energy supply chains in the two sub-nexi recognize that there are unacceptable risks in not promoting policies that take into account the essential mutuality of water security and energy security.

Concluding comments on the nexus approach

Water and energy are essential natural resources. They sustain societies, economies and ecosystems. Scientists and specialist professionals such as engineers have been tempted to deconstruct complex natural systems and to study and theorize them in separate disciplines. Equally separate professions have established infrastructures to allocate and manage them.

In the late 1980s it was noted that water security and food security were very closely linked, but little progress has been made in theorizing or understanding the links in the past three decades.

As a consequence, the current generation of scientists, legislators, and those managing private-sector supply chains have a special responsibility. They find themselves coping with the second failure of capitalism mainly brought about by assuming that natural resources come free and that there are no consequences of misuse. The first failure of capitalism became evident at the end of the eighteenth century at a time when capitalists believed that labour came free. Slavery was widespread, and other labour did not need to be considered fairly. These lethal and unsustainable assumptions were progressively eroded. Social democratic movements saved capitalism by forcing it to treat labour more fairly and install infrastructure and welfare systems that made the evolving neoliberal market systems potentially sustainable. Labour had a collective voice that made the risks of volatile politics and revolution very evident.

Coping with the second failure of capitalism is proving to have its own particular challenges. Natural systems have no collective voice. The risks associated with the ways that private-sector supply chains manage natural resources are incremental and can be invisible for decades. There is no capacity to consider such risks when the costs of adopting environmentally sustainable measures seem to be counter-intuitive. They would be costly. And they have to compete with more urgent investment priorities and the interests of shareholders and employees. In the case of the food supply chain, where a significant proportion of food supplies have to be cheap, the commercially weak farmers who manage food production and 92% of the water needed by society face impossible challenges. It is time to find a way to value natural resources, and the ecosystems that they provide, as crucial elements in sustainably stewarded global food and energy systems.

Scope of this *special issue* on global drylands

A climatically defined region of major concern is often under-represented in global nexus debates. Drylands host one-third of the global population, cover 41% of the land surface and are faced with greater challenges with respect to water, energy and food than other regions. Physical water scarcity is either already a major problem or may be exacerbated by climate change and over-depletion of available water resources. Hence, the nexus highlighted by the World Economic Forum is a topic that will define the twenty-first century, and not only in the respective dryland regions. In a globalized economy, the interlinkages through trade and climate change are evident. Dryland countries could be adversely affected by poor access to water, energy and food.

However, the picture is not all negative. On the one hand, drylands have untapped opportunities that need to be capitalized through cooperation. Science and technology should play a major and more prominent role as a basis of advice to decision makers charged with natural resources management. For example, the drylands have a considerable potential in increasing their use of water stored in the soil profile. At present, 90% of soil water in drylands is lost to evaporation (International Fund for Agricultural Development [IFAD], 2000). At a time of global economic change, increased pressure on the global food system and growing populations, this is one area where nexus thinking can provide a fresh approach to mitigation policies. Another area of importance is the role of energy. Since only some parts of dryland regions – such as in the Gulf Arab states – are well-endowed with fossil fuels, alternatives such as renewable energy could provide "win-win scenarios" to decrease water use and increase resources efficiency. These so-called "trade-offs" will have to be understood to increase water, energy and food security. At the same time, the policy environment will have to improve to induce reform measures that promote natural resources security.

Purpose of this special issue

The contributions to this special issue were peer-reviewed presentations to the International "Conference on the Water-Energy-Food Nexus in Drylands," held in Rabat, Morocco (the seat of the OCP Policy Center) 11 – 13 June 2014. Thanks to the generous funding of the OCP Policy Center, 30 researchers from across the Arab world, Europe and Africa discussed the nexus in a highly engaging format. All these contributors to improving ecosystem and natural resources management in global drylands emphasized the role of farmers. The conference paid particular attention to the role of policy to bridge the gap between science and policy. The nexus concept may hold a number of promises for dryland economies if its scientific recommendations are fully understood and sensibly integrated into policy. However, this will not be an easy task for decision makers in the years ahead. This special issue is a first contribution to the debate that will hopefully inform policy makers of opportunities related to the World Economic Forum's nexus concept. More importantly, it seeks to launch a debate on the concept and the future of natural resources management in dryland economies. It is hoped by the editors that scientists and policy makers will be able to grasp the nexus concept not only to manage resources in an integrated way but also to see the need to cooperate across borders and private-sector supply chains to solve the environmental challenges of the future.

Disclosure statement

No potential conflict of interest was reported by the authors.

References

Abu-Zeid, M. (2001). Water pricing in irrigated agriculture. *International Journal of Water Resources Development, 17,* 527–538. doi:10.1080/07900620120094109
Allan, T. (2011). *The water, energy and food security,* Interview at the Federal Government of Germany conference in Bonn – 16–18 November 2011. Retrieved from http://www.water-energy-food.org/en/news/view__225/tony-allan-the-tension-between-sustainability-and-intensification-captures-totally-what-this-meeting-is-about.html
Biswas, A. K., & Tortajada, C. (2010). Future water governance: Problems and perspectives. *International Journal of Water Resources Development, 26,* 129–139. doi:10.1080/07900627.2010.488853

British Petroleum. (2013). *Water in the energy industry: An introduction.* London. http://www.bp.com/content/dam/bp/pdf/sustainability/group-reports/BP-ESC-water-handbook.pdf

Falkenmark, M., & Rockström, J. (2004). *Balancing water for humans and nature: The new approach in ecohydrology.* London: Earthscan.

Famigletti, J. S. (2014). The global groundwater crisis. *Nature Climate Change, 4*, 945–948.

Gilmont, M. (2014). *Water governance: The politics of the emergence of reflexive-discursive water policy in California, Australia and Israel.* (Doctoral thesis). King's College London.

Hoff, H. (2011). Understanding the nexus. In *Background paper for the Bonn conference on the water-food-energy nexus.* Stockholm: SEI.

Hoogeveen, J. (2015). Consumptive use of water in agriculture, industry and households. E-Mail communication on 5 March 2015.

Hoogeveen, J., Faurès, J.-M., Peiser, L., Burke, J., & van de Giesen, N. (2015). GlobWat – a global water balance model to assess water use in irrigated agriculture. *Hydrology and Earth System Sciences Discussions, 12*, 801–838. doi:10.5194/hessd-12-801-2015

International Fund for Agricultural Development. (2000). *Sustainable Livelihoods in the Drylands: A discussion paper for the eighth session of the Commission on Sustainable Development.* Rome: Author.

Mekonnen, M. M., & Hoekstra, A. Y. (2011). *National water footprint accounts: The green, blue and grey water footprint of production and consumption.* Value of Water Research Report Series No. 50. Enschede: UNESCO-IHE.

Organisation for Economic Cooperation and Development. (2012). *Environmental outlook to 2050: The consequences of inaction.* Paris: Author.

Paarlberg, R. (2013). *Food politics.* New York and Oxford: Oxford University Press.

Pascual, C., & Elkind, J. (2009). *Energy security: Economics, politics, strategies, and implications.* Washington, DC: Brookings Institution.

Thompson, M. (1990). *Cultural theory.* Boulder: Westview Press.

Waughray, D. (Ed.). (2011). *Water security: The water-energy-food-climate nexus.* Washington and London: Island Press.

Financial challenges of the nexus: pathways for investment in water, energy and agriculture in the Arab world

Martin Keulertz[a] and Eckart Woertz[b]

[a]Department for Agricultural and Biological Engineering, Purdue University, West Lafayette, IN, USA; [b]Barcelona Centre for International Affairs, Barcelona, Spain

The Water–Energy–Food (WEF) nexus is a development challenge in the Arab world, particularly in the 'core nexus countries' with low to mid-incomes in which limited water endowments permit agricultural production, such as Egypt, Morocco, Tunisia, Lebanon, Algeria, Sudan and Jordan. The WEF nexus is often conceptualized in mere technocratic terms, yet politics matter in the implementation of projects that address it. Internalizing hydrological externalities or leaving them as they are and financing them as a public good requires states whose capacities have been reduced as a result of neoliberal reform. The article explores five different pathways of how Arab countries could finance green growth projects ranging from regional financial markets to concessionary loans by funds from oil rich Gulf countries.

Introduction

The Water–Energy–Food (WEF) nexus is one of the most crucial developmental challenges for semi-arid and arid economies such as in the Arab world. Food production is compromised by water scarcity; water scarcity in turn is caused by population growth, urban water supply and sanitation and agriculture (Allan, 2002). Other trade-offs exist between water and energy production and between energy and food production. As the Arab world's dependence on food, and therefore on virtual water 'imports' continues to grows, achieving water, energy and food security *sustainably* will be a challenge of historical magnitude. International organizations, the scientific and non-governmental organization (NGO) communities have repeatedly pointed at practices aimed at sustainable development in dryland economies. The emphasis is placed on policy shifts to generate positive trade-offs through resource optimization. According to those views, 'win–win scenarios' between economic and environmental planning can be achieved by following advice, from both the physical and the social sciences, that emphasizes planning tools for 'nexus implementation' (Hoff, Mohtar, & Lahn, 2014). While the rationale of those approaches is to identify demand-side management solutions, this paper will argue that these discussions are often incomplete. They neglect the pivotal point that demand-side management options will not suffice due to population growth, uneven economic development and past political economic spectres that continue to haunt Arab drylands. This article will therefore address the question of how to fund sustainable strategies that are informed by nexus thinking for both demand and supply-side management.

The article is structured in four parts. It will first provide a brief sketch of the history of the WEF nexus and why drylands are of paramount importance for global political and economic stability. Second, it will illustrate the challenges of nexus implementation strategies due to the lack of integration of political economic issues into them. Third, it will address state capacities in Arab dryland countries, which in part have been affected by neoliberal reform prescriptions in the 1980s and 1990s. These economic reforms have cast a long shadow on a crucial actor for the success of nexus approaches in the region: the state. Fourth, the article will recommend strategies on how to enable Arab countries with means to implement the nexus: strategies that are far greater than mere economic incentives and policy changes. It will be argued that implementation of nexus policies in Arab drylands requires a paradigm shift toward internalization of environmental costs in Arab economies and/or promoting the environmental benefits of sustainable water management as a public good.

Brief summary of the nexus and challenges in drylands

The first aim of this paper is to provide a definition and history of the WEF nexus, which has become a buzzword in global development used to highlight the systemic interlinkages between the water, energy and food sectors. Similar to the partly criticized water paradigm of the 1990s Integrated Water Resources Management (IWRM), the WEF nexus shares problems of measurement, operationalization and implementation that post significant challenges to decision-makers (Biswas, 2008; Giordano & Shah, 2014). Although academics highlighted the interlinkages for a long time (Lofman, Petersen, & Bower, 2002), the birthplace of the WEF nexus as a concept in development was the World Economic Forum, which investigated at its annual conference in 2008 the future role of water security in the global economy. Multinationals such as Nestlé, Coca-Cola and Dow Chemical provided financial support to consult leading international scientists on how to steward water resources into the 21st century, with the result of rediscovery of an approach based on systems theory. In the most general sense, system means a configuration of parts connected and joined together by a web of relationships (Banathy, 1996). The WEF nexus is based on systems thinking, which means it does not view the three sectors independently, but takes the perspective that all three sectors should be governed together. This means that the nexus encompasses the full usage of natural resources to produce water, energy and food. In particular, the crucial role of water for economic and environmental sustainability is echoed in nexus approaches and is the single most important contribution to the WEF nexus concept (Waughray, 2011).

The emphasis on water stems from its increasing importance over the past decades because water has become a limiting resource for economic growth. Blue water is finite (Holmgren & Berggren, 2014). However, in theory, the number can be expanded through costly investments in reuse and desalination plants, which is economically and environmentally costly due to it requiring substantial inputs. For developing countries, this is mostly unaffordable. Thus, humans will have to manage readily available water resources more efficiently to match the increasing demand for food and energy that is caused by population and economic growth.

Climate change research has added to the significance of water in current debates. Climate change may impact precipitation or, even more importantly, the reliability of rainfall. This means that the quantity of water available for food and energy production may alter over the coming decades, resulting in increased pressures on all three systems. Proponents of the WEF nexus rightly claim that the main focus of policy-making should be on the management of a country's available natural resources (Hoff, 2011; Wong, 2010). Addressing the nexus must find its way into policy and an improved collaboration between ministries and agencies in charge of the water, energy and food sectors. In practice, integrated policy-making could alleviate pressures on all three sectors by harmonizing the management of water, energy and food. For example, energy development plans should take the impact of proposed projects on the food and water sectors into account to avoid negative consequences on these sectors. In particular, nexus approaches could point policy-making towards new technologies in all three sectors that can decrease the use of water and free resources through improved efficiency in management. It is further suggested that water-scarce countries possess an enormous potential to benefit from these trade-offs by focusing on renewable energy and an optimization of 'green water' management (water in the soil profile).

Yet, despite the importance of integrated decision-making approaches, thought must also be given to how to implement the WEF nexus concept. German development cooperation has played a prominent role in introducing the WEF nexus concept into development debates and discussing its implementation. A major milestone of this process was the Bonn Nexus Conference of 2011 (Hoff, 2011). Yet, while the concept was generally welcomed, to date neither industrialized nor developing countries have integrated WEF nexus suggestions systematically into policy-making, although there have been attempts at bureaucratic unification, Morocco, for example, has integrated the ministries of energy, mining, water and the environment into a single ministry. One reason for this lack of implementation is the unclear role of investment and how these might achieve a greater harmonization of water, energy and food policies. This is the point of departure of this article. Governments in charge are often subject to severe fiscal pressures that have left their historical toll on dryland economies. We argue that WEF nexus policy-making is, by nature, a core aspect of statecraft. The world is not run by resource managers and engineers, but rather by politicians who are subject to global as well as to local influences. These may include multinational corporations, financial sectors, state politics, lobbyism, economic cronyism, tribalism, etc. and it is these various influences that form the 21st-century 'state'. By shedding light on the role of the 'state' in 'getting the nexus right', we argue here that geopolitical and political–economic contexts are crucial for the implementation of nexus-related policy approaches.

The geopolitical side

The Arab region is the most food trade-dependent region in the world. As Allan stresses, 'The Middle East has effectively ran out of water resources to become self-sufficient through domestic food production since the 1970s' (Allan, 2002, p. 9). As a result, approximately 50% of all food consumed in the Arab world is imported from regions that have sufficient water to produce beyond their domestic requirements (World Bank, 2008). In 2010, the Arab region was the largest importer of cereal grains, at 65.8 million metric tonnes (Solh, 2013). Thus, Arab economies are prime customers of traders of agricultural staple commodities such as Archer Daniels Midlands, Bunge, Cargill, Louis Dreyfus and Glencore (*The Economist*, 2013). Ongoing global economic change makes this very problematic. Asian countries have become more prosperous and are shifting their dietary preferences towards meat and dairy products. Not all of this demand can be satisfied by domestic production, making China in particular severely water stressed and causing it increasingly to resort to importing feedstock, particularly soybeans, for its livestock industries, from the US and Brazil. China continues to maintain a self-sufficiency policy for grains for human consumption, but has started to rely on some supplemental imports as well. Additionally, it procures large amounts of industrial agricultural raw materials from abroad, such as rubber or timber. To satisfy these import needs China has fostered trade relations and developed strategies to invest in agricultural value chains in Latin America, North America, Eastern Europe and, to a lesser extent, Africa. In the 'good old days', the Arab region was strategically of high importance, while Western agricultural centres, suffering from overproduction, had interest to place their agricultural surpluses at subsidised rates on markets in the Eastern Bloc and developing world. Asia's increasing appetite for food and the 'virtual water' embedded in such food, will increasingly affect this geopolitical Equation (Woertz & Keulertz, forthcoming).

It has become brittle for other reasons as well. The geopolitical importance of Arab economies increased during the second part of the 20th century. The advent of the 'carbon age' (Mitchell, 2011) and the founding of Israel gave the Arab region a special place in world politics. A superpower rally to curry favour with regimes emerged throughout the Cold War period. In this context, the US politicized the food trade, using it to achieve foreign policy goals. Subsidised food aid deliveries under the PL-480 programme were granted or withdrawn to entice cooperation from recipients like Egypt, Jordan and Syria. During the time of the Arab oil boycott, even a retaliatory food embargo was contemplated (Woertz, 2013). The dependence on food imports from regions with readily available water resources has been perceived as a strategic vulnerability by Arab governments ever since. Such concerns were reinforced with the commodity food price spikes of 2007/08 and 2010/11. There is growing awareness that the global food market may not absorb increasing demand for food imports as easily as in the past, right at a time when Arab countries fear that their geopolitical importance has diminished. The US now puts greater weight on the Pacific region in its foreign policy. While the Gulf Arab countries will always be crucial for price discovery of a fungible commodity like oil on a unified global market, America's perceived reliance on Gulf oil supplies has decreased as a result of its unconventional oil and gas revolution via hydraulic fracturing. On the other hand, the food trade has been depoliticized since the 1980s and has reoriented itself toward commercial considerations, including a reduction of export subsidies under the World Trade Organization (WTO) umbrella and a shift of food aid from bilateral to multilateral distribution channels. This could affect the availability of subsidised food imports for the poorer Arab countries.

The strategic importance of the Arab region allowed its policy-makers to bet on the food import option, reducing the need to address the optimization of the use of water, energy and food. The changing geopolitical context means that nexus challenges will need to be addressed and will require significant investment in infrastructure, capacity and governance – both financially and politically.

The investment challenge of the nexus

Optimizing the use of water, energy and food resources affects supply and demand measures and related investments. The United Nations Economic and Social Commission for West Asia (UNESCWA) suggests that the combined investment required for 'green growth' demands an investment of 2.3% of the region's gross domestic product (GDP) (UNESCWA, 2013). 'Green growth' is an umbrella term that describes a path of economic growth that ensures the management of natural resources in a sustainable manner. Conceived by international organizations such as the United Nations, the World Bank and Organisation for Economic Co-operation and Development (OECD) and governments like

South Korea, it seeks to devise an alternative economic strategy at a time of increasing pressures on natural resources. Water, energy and food are three of the most crucial sectors for achieving green growth due to their particular role in natural resources management (UNESCWA, 2013).

Translated into the fiscal year 2013, 'green growth' strategies would mean investment costs of approximately US$82 billion per annum across the Middle East and North African (MENA) region (UNESCWA, 2013). The region's uneven economic development means that more than half of the MENA region's GDP is generated in the oil-rich Gulf countries. This distortion needs to be taken into consideration. More importantly, the hyper-arid nature of the Gulf significantly increases investment needs in the water sector.

Breaking it down into the three sectors (water, energy and food) that are particularly important for green growth, the energy sector is estimated to require an approximately US $316 billion investment in upstream, downstream and power generation by the end of this decade to cater to the future needs of Arab populations (APICORP, 2014). The bulk of this estimate relates to hydrocarbon energy. However, renewable energies play an increasing role in the planning of Arab countries. Some areas, like the Atlantic coast of Morocco, have favourable conditions for wind energy and prices for photovoltaic cells have come down dramatically. The future development of storage solutions could ease the integration of renewable energies into electricity grids. A more balanced energy mix with the inclusion of renewable energy could provide low carbon water generation including desalination, water treatment and reuse. Finally, off-grid energy production could promote rural electrification, water pumping, irrigation and low carbon food production.

For the agricultural sector, Arab countries are the largest net importers of grains in the world; approximately 56% of their consumed cereal calories are imported (Al Masah, 2012). Grain self-sufficiency is economically very costly, if not impossible due to water constraints (Magnan, Lybbert, McCalla, & Lampietti, 2011). Some politicians estimate the investment costs to increase regional agricultural production at US$80 billion until 2020 (Arab News, 2013). However, those estimates need to be treated with extreme caution due to the unclear role of global food trade. If world commodity prices remain stable, it will be much cheaper to continue to import food than investing in regional agriculture.

For the water sector, cumulative annual investment needs may vary dramatically in relation to climate change projections, from US$27 billion (wet conditions) to US$212 billion (dry projections) until the year 2050 (Immerzeel et al., 2011). Thus, investment in water infrastructure could significantly outstrip energy and agricultural investment costs, turning it into the great unknown variable of future investment needs of the Arab world. Nexus thinking can offer fresh perspectives on these investment needs. In dry, marginal areas, about 90% of rainwater is lost to evaporation (Cleveringa, Kay, & Cohen, 2009). Making use of this green water potential in spite of climate change could relieve investment pressures on the water and agricultural sectors. Rainfed dryland agriculture with supplemental irrigation could reduce spending on irrigation infrastructure and also on desalination, which is mainly used for residential supplies, but sometimes also for landscaping and agriculture (e.g. in the Gulf region). At the same time, the energy sector could provide sustainable inputs to both systems, if engineered by renewables. All sectors will, nevertheless, require significant public investment to meet the targets envisaged by UNESCWA.

Investment will need to target physical infrastructure for water, energy and food provision, but will also need to support farmers as they optimize soil and water use. This will require greater investment in agricultural research and development (R&D). R&D activities involve research in the general sciences, the pre-technology sciences and technology invention. Innovations from such activities result in products that can be extended to final users, such as farmers, consumers and government agencies (King, Toole, & Fuglie, 2012).

Due to the role of agricultural R&D as a public domain, it is not easily privatized. Thus, the state will need to play a significant role in technology transfer and agricultural extension to allow farmers access to latest agricultural research (King et al., 2012). Institutions for agricultural R&D in Arab countries require substantial improvement: Egypt's National Agricultural Research Centre, for example, employs approximately 100,000 people, but the extension services to support farmers is very weak. Incorporating the nexus into agricultural research centres, such as in Egypt, requires a strengthening of the scientist–farmer exchange and related technology transfer (Piotrowski, 2014).

Farmers will also require financial assistance through credits and crop insurance. Lastly, improved supply chain management will require better rural infrastructure such as transportation to connect rural areas with urban markets, cooling chains, storage and an up-scaling of farmer cooperatives. These investment requirements apply to the majority of Arab dryland countries. The nature of the nexus challenges affects wider socio-political concerns. It will require cooperation between the public and private sector as well as civil society. However, in order to steer and manage investment in water, energy and food, the state will have to play a crucial role, bringing us to the political–economic side of the nexus.

State of affairs in the Arab public sector

The role of the state will be crucial in preparing the region for WEF nexus challenges and spur-related investment. However, uneven resource endowments have led to uneven regional economic development. This translates into very different challenges for nexus implementation. We therefore classify the region into three categories.

(1) Core nexus countries: low to mid-income countries in which some water endowment permits agricultural production. Agriculture has played a traditionally important role in countries like Egypt, Morocco, Tunisia, Lebanon, Algeria, Sudan and Jordan. Investment in the agricultural sectors of these economies could substantially increase regional food security. At the same time, these economies require increased investment in the energy and water sectors to provide growing populations with increased access to water, food and energy. In these economies, all three segments of the WEF nexus need to be addressed to secure economic growth and resource security.

(2) Non-core nexus countries: high-income countries with arid climates in which agriculture will be a limited option due to the lack of water. These countries are mainly located in the Persian Gulf region, and include Saudi Arabia, Oman, Qatar, Kuwait, Bahrain and United Arab Emirates. In the non-core nexus countries, the main challenges are related to achieving water and energy security domestically. While food security is of paramount concern, physical water scarcity and climate change render these countries less promising for domestic food production. While some have created a large subsidised agricultural sector, like Saudi Arabia, this sector must now be cut back because fossil water aquifers have been depleted. The food import option will remain their primary choice to provide food security. As a result, in the non-core nexus countries, the water and energy nexus is of main concern but not the full WEF nexus.

(3) Countries exposed to internal or external conflict outside of the scope of a nexus analysis: although countries like Syria, Yemen, Palestine, Iraq and Libya are partly blessed with promising agricultural opportunities, the negative impacts of conflict on their current economic development render a nexus-related analysis difficult at the present stage.

The core nexus countries are also those in which the public sector is subject to severe pressures due to limited hydrocarbon endowments, lack of economic development and governance issues. It is argued here that these are the countries of major concern for the implementation of the nexus due to their potential to stabilize food production in the face of growing populations.

Apart from Jordan and Lebanon, all the core nexus countries share a common economic history in the period after the Second World War. The widespread adoption of Import Substitution industrialization (ISI) policies focused financial resources in the hands of the state via nationalization, granted subsidies and tariff protection to industry sectors it deemed strategic, and fostered a new bureaucratic urban middle class via increased state spending. However, the fledgling industries subsidised never matured and were in constant need of support and foreign exchange. Import dependence merely shifted from manufactured goods to investment goods, proving unsustainable and leading to the debt crisis of the 1980s in vast parts of the developing world, including the majority of core nexus countries of the Arab region.

The involvement of the International Monetary Fund (IMF) and the World Bank led to structural adjustment programmes in exchange for loans. Those loans were widely administered during the period 1985–95 and in the years after 2003. The largest beneficiaries were Jordan, Algeria and Egypt (1985–95), and Iraq, Morocco, Sudan and Jordan (after 2003). Welfare provision and state subsidies were cut, among them those for agriculture (Henry & Springborg, 2010). In pure metrics, public finances in the region improved. Austerity policies led to reduced spending, but direct and indirect rents from oil and gas, the region's main natural endowments, also helped lower public debt. As a result, public gross debt in most Arab countries has decreased since the 1990s (IMF, 2014b). With the exception of Lebanon, debt levels are well below those of OECD countries, which often have public debt levels of 100% and more. The Gulf countries never had to turn to the Bretton Woods institutions for assistance. They have considerable fiscal flexibility and have reduced their debt levels to near zero in the 2000s. However, they remain vulnerable to shocks, as rising spending and domestic oil consumption constantly pushed up the oil price ceilings required to balance their budgets.

The underlying principles of neoliberalization and the Washington Consensus were only partly implemented as conceived by economic advisors in Washington, DC. Private sector growth remained limited because this would have interfered with the power structures in the respective political economies. The shadow states of powerful cronies were not touched (Malik & Awadallah, 2011; Yorke, 2013). The opportunities of foreign direct investment (FDI) were never fully utilized in the core nexus countries and the Arab region remains punching below its weight in the global economy (Henry & Springborg, 2010; Richards & Waterbury, 2008).

Most crucially in relation to the WEF nexus, austerity policies without a strong capital inflow alternative and facing population growth led to problems in the public sector to governing water, energy and food effectively. Consequently, infrastructure degradation and neglect of basic public services, such as education and R&D, impacted the provision of core services that WEF nexus policies require. At the same time, poverty levels often did not decrease because the expected 'trickle down' effect did not occur. This left core nexus countries like Egypt, Jordan and Morocco with underlying structural poverty risks that have not vanished to date (El-Said & Harrigan, 2014).

In sum, the financial position of Arab governments with food production potential improved as a result of austerity measures and, in some cases, higher oil prices. Yet, since populations continuously grew, more money would have been needed to provide public services. This led to an investment gap. Investment in water, energy and food will be crucial to provide basic services to populations, but the state in the core nexus Arab countries will hardly be able to meet the investment needs unless financial alternatives are made available. A core focus of the neoliberal paradigm is placed upon private sector development. Yet, it is in the nature of private capital to invest in goods and services that provide financial returns. While the energy and agricultural sectors could, in theory, provide financial returns for private capital, water is a sector with distinct externalities because it cannot be easily priced. Internalizing the hydrological externalities or leaving them as they are and financing them as a public good will be the core challenge of WEF nexus approaches.

Externalities and public goods – the role of the public sector

In the water debate, it is disputed whether water is a private or a public good. Although some describe water as an economic good that should be privatized to enable a more efficient use of (blue) water resources (Rogers, Silva, & Bhatia, 2002). Such debates have bias toward urban 'blue water' use/supply (Savenije, 2002). While we acknowledge the possibility to leave urban water *supply* to the private sector, the discussion of whether water *itself* is a private good is more complex.

Water is different from other goods as a result of its various characteristics like bulkiness, low tradability and heterogeneous demand patterns that span vastly different categories like residents, industry, farmers and those who are unable to pay (e.g. the environment, poor people). Water has limited substitutability and its essential importance for human life leads to low excludability and preponderance of states in its management. These caveats become even more important when one goes beyond the limited realm of localized drinking and irrigation water supplies and takes a holistic view of the water resources sector. In particular, this applies to semi-arid and arid regions, where water is an increasingly scarce resource.

In neo-classical economics, the most apt way to deal with a scarce economic resource and achieve equilibrium is the market. Positive or negative externalities of such market transactions that affect third parties that have not been part of the initial transaction constitute market failures that can be corrected by regulatory government action, ideally by internalizing external effects through pricing mechanisms (OECD, 2003). However, internalizing water costs is a difficult task. The WEF nexus deliberately includes blue (surface and ground water) and green water (water in the soil profile) in order to take all available water resources into account (Hoff, 2011). As a rule of thumb, 70% of global water resources are used in agriculture; 20% by industry and only 10% for domestic usage (Hoff, 2011). While the 20% by industry and 10% by domestic water users can be recycled through wastewater treatment, agriculture takes the largest toll on consumptive use of water resources with 90% (Hoff, 2011; Rockström et al., 2009). Hence, the bulk of water within the nexus is used by the agricultural sector, which Allan (2013) calls 'food water'.

Water for food production is subject to characteristics that are difficult to price: it can only be used and accessed, but not actually owned. 'Water is a fugitive common property resource' because if understood through the hydrological cycle, water is hard to capture for private ownership (Agarwala & Allan, 2014). Water falls on earth through rain, moves in rivers, lakes and streams, seeps through the soil to recharge groundwater aquifers, and evaporates back into the atmosphere through the hydrological cycle. While the hydrological cycle means that water turns into vapour that returns to the Earth's surface as more rain, it cannot be predicted where or when it will do so. Water is unevenly distributed around the globe, according to climate zones and the availability of ground and surface water. The international community has declared water to be a human right. All these are strong factors that inhibit pricing water: it is a local resource, costly to move, and best used in regions with water availability (He, Tyner, Doukkali, & SIAM, 2006; Holmgren & Berggren, 2014; Molle, 2008). Finally, water is deeply connected with cultural factors and traditionally has been free throughout human history; hence it never had to be taken into account for pricing food products (Allan, 2013).

If water for food production has the character of a public good with externalities that cannot be easily internalized, the question arises as to who provides the public good and the necessary financial means to maintain it. From a Marxist perspective it has been argued that this represents a basic contradiction of capitalism, which subjugates limited natural resources to a limitless logic of accumulation, by relying on a constant externalization of costs to entities operating outside its logic (nature, reproduction of labour via families, unpaid female household work etc.) and a privatization of public goods and infrastructure provided by the state, without which it cannot operate in the long run, but which it requires in the short run to cope with crises via 'accumulation by dispossession' (Harvey, 2003). A sustainable management of resources may well prove impossible on this ground. Yet, on a more 'reformist' plane, one may ask whether a more balanced management of externalities could be achieved by fostering nexus projects such as increased investment in both green and blue water management. This invariably raises the question who could provide the finance for such projects?

Financing the nexus: alternative pathways?

We identify five alternative pathways to finance projects related to the water, energy and food sector. These policy options involve international, regional and domestic financial support for green growth projects. The first option is to make use of climate finance instruments, a financial concept that comprises conditional loans from industrialized countries in support of climate change mitigation and adaptation in developing countries. At the United Nations Framework Convention on Climate Change (UNFCCC) Conference of the Parties in Cancun (COP16), industrialized countries committed to mobilize US$100 billion per year by 2020 to provide grants, grant components of concessional loans and public support from public and private, bilateral and multilateral sources to support less affluent countries. Partly financed through emissions trade, the overall goal is to minimize CO2 emissions globally through a trade system that allows industrialized countries to buy off emission rights from developing countries (GIZ, 2014). However, despite grand ministerial announcements, the available funds are not necessarily new, but often part of the Official Development Assistance (ODA) portfolio of developed countries. For example, the new Green Climate Fund of the United Nations has raised US$10 billion (with a target to increase it to US$100 billion by 2020) in order to support poor nations in coping with climate change. However, the committed loan opportunities, thus far, lag behind actual finance requirements of developing countries. Although these loans are available to Arab governments, they mark a drop in the ocean, given the costly investment needs of the region.

The second option is domestic or regional capital to fund investment requirements. As Table 1 illustrates, banks dominate the capital structure of Arab capital markets with 62%. Equity markets play a relatively minor role in global comparison, with 29%; bond markets are completely underdeveloped, with only 7% of overall financing (Table 1). A selected minority of large companies and consumer financing has been the main beneficiary of the considerable credit growth in the region. Credit allocation has been dependent on proximity to political power. The financial sector and the real private economy, on the other hand, have seen a certain disconnect. It has been difficult for small and medium-sized enterprises (SMEs) in family ownership to obtain bank credit and they have been reluctant to enter the stock market, relying instead on retained earnings (IMF, 2014a; World Bank, 2006).

As a remedy, it has been suggested to improve the access of SMEs to credit and upgrading the capacities of specialized funding vehicles such as agricultural investment banks. The (partial) privatization of state-owned enterprises, initial public offerings (IPOs) of family companies and a strengthening of an institutional investor base could increase liquidity and free floats on domestic stock markets. The development of corporate bond markets would require the issuance of government bonds to form a benchmark (Woertz, 2012).

Yet, it is unclear whether nexus projects could obtain funding from such capital markets, even if their structure and accessibility were improved. Nexus projects aimed at improving water, energy and food management could do so only if they turned a profit. In the energy sector, this often will only be possible with future technological progress. Storage solutions for intermittent renewable energy sources, for example, could ease their integration into

Table 1. Stock market capitalization, total debt securities and bank assets in the Middle East and North African (MENA) region (IMF, 2014a).

	US$ billions	Percentage of gross domestic product (GDP)
Stock market capitalization		
Middle East and North Africa	895	29
Latin America and the Caribbean	2476	44
Emerging markets Asia	5853	47
North America	18,883	105
European Union	10,086	65
World	52,848	73
Total debt securities, 2012		
Middle East and North Africa	221	7
Latin America and the Caribbean	3590	64
Emerging Markets Asia	5492	44
North America	37,292	206
European Union	29,297	189
World	98,974	137
Data include total debt securities, all issuers and amounts outstanding by residence of issuer		
Bank assets, 2012		
Middle East and North Africa	1921	62
Latin America and the Caribbean	3948	70
Emerging Markets Asia	21,081	170
North America	18,679	103
European Union	47,856	308
World	121,947	169
Data include total assets of domestic commercial banks, including foreign banks' subsidiaries operated domestically		

electricity grids decisively, further reducing costs and make them more competitive with carbon energy sources. At an early stage, renewable energies require prolonged subsidies and soft finance that only governments are willing and able to provide. The Internet was subsidised over decades by the US government before it had sufficiently matured to provide a self-sustained platform for profit-oriented private sector capital. Another aspect of nexus projects' profitability is that they often require regulatory steps to compensate them for their provision of public goods and their reduced negative externalities compared with conventional economic activities. However, the underlying resource for economic and capital growth is water. Thus, getting the nexus right translates into the right investment in water management from infrastructure to re-education of farmers.

As rentier and semi-rentier states in the Arab world have relatively low tax rates compared with most OECD countries, their ability is limited to provide incentives and sanctions via this avenue to fund nexus projects. On the other hand, they entertain extensive subsidies for fuel, electricity and food. A redirection of such subsidies from hydrocarbon sources to renewable energies could provide a development impetus for nexus infrastructure and R&D projects. Yet, it would require awareness that change is inevitable by the beneficiaries of present hydrocarbon subsidies. Successful development in this direction would require an active developmental state to introduce bold subsidies reform such as channelling current hydrocarbon subsidies into renewables. The political–economic rationale of both rentier economies and 'bunker states' serves different political allocation priorities of their bureaucratic constituencies (Henry & Springborg, 2010). Thus, the policy environment is crucial to understanding the implementation challenges of the WEF nexus. Policy-makers must understand the nexus in order to support greater attention toward the three nexus systems through integration of investment and management policies by all ministries charged with natural resources. This addresses not only the ministerial level but also, indeed, the highest level of decision-making, e.g. royal families, presidents or prime ministers; moreover, it applies to both sides: governments in need of investment capital and governments with available funds for loans, etc.

The third option would be to provide development banks with increased access to capital to act as intermediaries between the public and private sector. Yet for the same reasons as above, the fiscal space and bureaucratic capacities of non-oil-exporting Arab governments to provide nexus funding via their specialized national funding vehicles is limited. In fact, the development finance sector has been in recession since 2008. Regional and international banks that previously provided project finance and infrastructure deals in the Arab world declined from over 40 to well below 20 (Collins & Godfrey, 2014). A major reason for the disappearance of these banks has been the global financial crisis leading to a further withdrawal of international capital from the Arab region.

Meanwhile, the fourth available option is the raising of 'traditional' concessionary financing from abroad via international financial institutions and bilateral investment agreements. The World Bank, for example, provided a major (US$297 million) loan facility to finance the Ouarzazate Concentrated Solar Power Plant Project in Morocco (World Bank, 2011). However, this would leave the region dependent on the Bretton Woods institutions and there is a certain reluctance regarding the strings attached to such finance. These strings often include wider economic reforms that may have profound political–economic impacts, which may not be wanted by governments in charge of state affairs.

Thus, if the water, energy and food challenges are understood and the political will exists, the most viable option to finance the nexus agenda could be from the region itself from non-core, oil and gas-rich Gulf countries. This fifth option would mean a repetition of regional history. Potential funders from the non-core nexus countries could be development funds such as the Kuwait Fund for Arab Economic Development (KFAED), the Saudi Fund for Development (SFD), the Abu Dhabi Fund for Development (ADFD), the multilateral Arab Fund for Economic and Social Development that is based in Kuwait, the Khartoum-based Arab Bank for Economic Development in Africa (BADEA), the Jeddah-based Islamic Development Bank (IsDB), and the Vienna-based OPEC Fund for International Development (OFID). The latter two are not strictly Arab funds, but Gulf Cooperation Council (GCC) countries provide a large share of their funding, most notably Saudi Arabia and Kuwait (World Bank, 2010).

Gulf countries have enjoyed a reputation of generosity in global development since the 1970s. By international standards, Arab ODA was exceptionally high, with 1.5% of their GDPs on average. This was more than double the UN targets of 0.7% of GDP as promised by OECD governments in the Millennium Development Goals, which were never met by most OECD countries (World Bank, 2010). However, Gulf lending was particularly high during the 1970s, with 3.8% of GDP and 1.5% from 1980 to 1989. Since then it has fallen to the 0.7% target in the 1990s, and further to 0.5% in the 2000s. This is still considerably higher than the actual OECD-Development Assistance Committee average of 0.3%, but an increase would be conceivable given high oil prices, low debt levels and large foreign exchange reserves. Via their development funds, GCC countries have experience in co-financing projects of international donors and vetting associated investment proposals. They also coordinate their activities via the Coordination Secretariat of Arab National and Regional Development Institutions, which is based at the Arab Fund for Economic and Social Development in Kuwait. Such development funds would form a better base for a concerted effort at nexus financing by non-core nexus to core nexus countries than political distribution channels, if coordinated successfully with development institutions in target countries.

There are discussions whether all such transactions should be regarded as development aid. Often they are primarily motivated by political considerations. Especially after the Arab Spring, they have been used to buttress friendly regimes in the region. Allocation has often been via fragmented and competing ministerial bureaucracies. In order to remedy past fragmentation problems, the nexus could serve as an opportunity to channel capital directly into the water, energy and food systems: but it all depends to what extent the challenges of managing water, energy and food inextricably will be understood – and how soon such projects could give an impetus for regional economic and political stability.

Conclusions

The WEF nexus is a concept of major importance for the Arab world. It has gained the attention of governments, development agencies and the academy across the region. It is logically appealing, however: the world is not run or governed by resource managers and engineers but rather by individuals charged with state affairs and public finances. The WEF nexus cannot be addressed without taking the role of the state into account. In particular, water management can hardly be provided by the market alone because of the very nature of water as an externality. Providing a regulatory framework for internalizing negative externalities via sanctions or providing positive externalities as a public good can only be exercised by the public sector. However, the public sector is exposed to financial challenges in a region troubled by a history of uneven economic development and mismanagement of public finances. Thus, embedded in the challenges of the nexus is a financial challenge: neither the state nor the market can effectively provide solutions without alternative financing. While the private banking sector has severely shrunk due to the global financial crisis, funding from the private sector may remain unavailable for the foreseeable future. A potential remedy for nexus challenges could be a return to the practices of the 1970s and 1980s when the rentier states of the Gulf region provided generous amounts of capital as ODA to developing countries. Since there are hardly any alternative options available, nexus initiatives will have to be funded through similar measures in the coming years to avoid a resources crunch.

The real question for those working on the WEF nexus will be to what extent the value of water will be understood. Inertia is not an option given the dramatic geo-economic shifts with emerging powers in Asia. As a result, the nexus presents a topic for a renewed Arab solidarity that will have to be defined and developed over the coming years.

Disclosure statement

No potential conflict of interest was reported by the authors.

References

Agarwala, M., & (Tony) Allan, J. A. (2014). Sustainable development of water resources. In G. Atkinson, S. Dietz, E. Neumayer, & M. Agarwala (Eds.), *Handbook of sustainable development* (2nd ed.). Cheltenham: Edward Elgar.

Al Masah. (2012). *MENA food security: Are we doing enough to feed the population?* Dubai: Al Masah Capital.

Allan, J. A. (2002). *The Middle East water question*. London: I.B. Tauris.

Allan, J. A. (2013). The food–water value chain. In M. Antonelli & F. Greco (Eds.), *L'acqua che mangiamo: Cos'è l'acqua virtuale e come la consumiamo*. Milano: Edizione Ambiente.

APICORP. (2014). *MENA energy outlook*. Dammam: APICORP.

Arab News. (2013). Arab States need $80.65 bn in agricultural investment. *Arab News*. Retrieved April 3, from http://www.arabnews.com/news/446878 (accessed December 10, 2014).

Banathy, B. A. (1996). *Designing social systems in a changing world*. New York: Plenum Press.

Biswas, A. K (2008). Integrated water resources management: Is it working? *International Journal of Water Resources Development, 24*, 5–22. doi:10.1080/07900620701871718

Cleveringa, R., Kay, M., & Cohen, A. (Eds.). (2009). *InnoWat: Water, innovations, learning and rural livelihoods*. Rome: IFAD.

Collins, N., & Godfrey, M. (2014). *Privatisation & public private partnership review 2013/14* Euromoney yearbooks. Abu Dhabi: Latham & Watkins LLP.

El-Said, H., & Harrigan, J. (2014). Economic reform, social welfare, and instability: Jordan, Egypt, Morocco, and Tunisia, 1983–2004. *The Middle East Journal, 68*, 99–121. doi:10.3751/68.1.15

Giordano, M., & Shah, T. (2014). From IWRM back to integrated water resources management. *International Journal of Water Resources Development, 30*, 364–376. doi:10.1080/07900627.2013.851521

GIZ. (2014). *Climate finance for adaptation in the water sector in the MENA region*. Cairo: GIZ and ACCWaM.

Harvey, D. (2003). *The new imperialism*. Oxford: Oxford University Press.

He, L., Tyner, W. E., Doukkali, R., & SIAM, G. (2006). Policy options to improve water allocation efficiency: Analysis of Egypt and Morocco. *Water International, 31*, 320–337. doi:10.1080/02508060608691935

Henry, C., & Springborg, R. (2010). *Globalization and the politics of development in the Middle East*. Cambridge: Cambridge University Press.

Hoff, H. (2011). *Understanding the nexus. Background paper for the Bonn conference on the water-food-energy nexus*. Stockholm: SEI.

Hoff, H., Mohtar, R., & Lahn, G. (2014). The MENA nexus initiative for a green economy, human security and stability. Working paper for the Global Water Partnership Conference "Sustainability in the Water-Energy-Food Nexus" 19–20 May 2014, Bonn Germany. Unpublished.

Holmgren, T., & Berggren, J. (2014). SIWI: Water is precious; it's time to advance our thinking around it. *The New Economy*. Retrieved from http://www.theneweconomy.com/strategy/siwi-water-is-precious-its-time-to-advance-our-thinking-around-it (accessed December 10, 2014).

IMF. (2014a). *Global financial stability review*. Washington, DC: IMF.

IMF. (2014b). *IMF data mapper*. Washington, DC: IMF.

Immerzeel, W. W., Droogers, P., Terink, W., Hoogeveen, J., Hellegers, P., Bierkens, M., & van Beek, R. (2011). *Middle-East and Northern Africa water outlook. World Bank study*. FutureWater Report 98.

King, J., Toole, A., & Fuglie, K. (2012, September). Complementary roles of the public and private sectors in U.S. agricultural research and development, EB-19. *United States Department of Agriculture, Economic Research Service*.

Lofman, D., Petersen, M., & Bower, A. (2002). Water, energy and environment nexus: The California experience. *International Journal of Water Resources Development, 18*, 73–85. doi:10.1080/07900620220121666

Malik, A., & Awadallah, B. (2011). *The economics of the Arab spring* CSAE working paper WPS/2011-23. Oxford: Centre for the Study of African Economies.

Magnan, N., Lybbert, T. J., McCalla, A. F., & Lampietti, J. A. (2011). Modeling the limitations and implicit costs of cereal self-sufficiency: The case of Morocco. *Food Security, 3*, 49–60. doi:10.1007/s12571-010-0103-2

Mitchell, T. (2011). *Carbon democracy: Political power in the age of oil*. London: Verso Books.

Molle, F. (2008). *Can water pricing policies regulate irrigation use?* World water congress, Montpellier.

OECD. (2003). *Pricing water*. Paris: OECD. Retrieved from http://www.oecdobserver.org/news/archivestory.php/aid/939/Pricing_water.html (accessed May 23, 2014).

Piotrowski, J. (2014). Smallholder farmers remain left out of most R&D. *SciDevNet*.

Richards, A., & Waterbury, J. (2008). *A political economy of the Middle East*. Boulder, CO: Westview Press.

Rockström, J., Falkenmark, M., Karlberg, L., Hoff, H., Rost, S., & Gerten, D. (2009). Future water availability for global food production: The potential of green water for increasing resilience to global change. *Water Resources Research, 45*, W00A12.

Rogers, P., Silva, R. D., & Bhatia, R. (2002). Water is an economic good: How to use prices to promote equity, efficiency, and sustainability. *Water Policy, 4*(1), 1–17. doi:10.1016/S1366-7017(02)00004-1

Savenije, H. H. (2002). Why water is not an ordinary economic good, or why the girl is special. *Physics and Chemistry of the Earth, Parts A/B/C, 27*, 741–744. doi:10.1016/S1474-7065(02)00060-8

Solh, M. (2013). *The outlook for food security in the Middle East and North Africa*. Rosenberg International Forum on Water Policy, 2013. Aqaba, Jordan.

The Economis. (2013 March). By the receding waters of Babylon. *The Economist*.

UNESCWA. (2013). Green growth in the ESCWA region. In UNESCWA (Ed.), *Roadmap for a green economy in the ESCWA region*. Amman: UNESCWA.

Waughray, D. (2011). *Water security: The water-food-energy-climate nexus*. London: Earthscan.

Woertz, E. (Ed.). (2012). *GCC financial markets: The world's new money centers*. Berlin: Gerlach Press.

Woertz, E. (2013). *Oil for food: The global food crisis and the Middle East*. Oxford: Oxford University Press.

Woertz, E., & Keulertz, M. (forthcoming). States as actors in international agro-investments. *International Development Policy*.

Wong, J. L. (2010). The food-energy-water nexus. *Harvard Asia Quarterly, 12*, 15–19.

World Bank. (2006). *Middle East and North Africa, economic developments and prospects 2006: Financial markets in a new age of oil*. Washington, DC: World Bank.

World Bank. (2008). Brief MENA region. In World Bank (Ed.), *Agriculture and rural development*. Washington, DC: World Bank.

World Bank. (2010). *Arab development assistance: Four decades of cooperation*. Washington, DC: World Bank.

World Bank. (2011). *World Bank supports Morocco's bold solar power plans*. Washington, DC: World Bank.

Yorke, V. (2013). *Politics matter: Jordan's path to water security lies through political reforms and regional cooperation* NCCR working paper no. 2013/19. Geneva: NCCR.

Food-water security and virtual water trade in the Middle East and North Africa

Marta Antonelli[a] and Stefania Tamea[b]

[a]Department of Design and Planning in Complex Environment, University IUAV of Venice, Venice, Italy; [b]Department of Environment, Land and Infrastructure Engineering, Politecnico di Torino, Turin, Italy

The purpose of this study is to analyze the political economy of food-water security in the water-scarce Middle East and North Africa region. The study deploys the lens of virtual water trade to determine how the region's economies have met their rising food-water requirements over the past three decades. It is shown that the region's water and food security currently depend to a considerable extent on water from outside the region, 'embedded' in food imports and accessed through trade. The analysis includes blue (surface and groundwater) and green water resources.

Water and food security in MENA

The Middle East and North Africa (MENA) region is acknowledged as having the largest water deficit in the world (Gleick, 2000; World Bank, Mohamed, & Kremer, 2009). Per capita freshwater availability decreased from $4000 \, m^3/y$ in 1950 to $1100 \, m^3/y$ in 2007 (World Bank, 2007). The region is wide and heterogeneous and it includes both rentier and non-rentier economies, with different socio-economic adaptive capacities and environmental frameworks that determine to a large extent the water policy and management. Absolute and per capita blue water resources in the Middle East are the lowest in the world (Food and Agriculture Organization of the United Nations, 2003). Renewable water resource withdrawal in the region already exceeds the critical thresholds of 20% and 40% of total renewable water resources, and water tables are falling as farmers and cities abstract water faster than the rate of replenishment from recharge and aquifer leakage (Food and Agriculture Organization of the United Nations, 2011). Since the 1970s, the water demands of its political economies have exceeded the capacity of the local resources for food self-sufficiency (Allan, 1997). In MENA there is "virtually no more freshwater to develop" (Food and Agriculture Organization of the United Nations, 2000, p. 50).

Despite being one of the major environmental challenges currently facing MENA, water scarcity is rarely taken into consideration when formulating economic policies in the region (Sakmar, Wackernagel, Galli, & Moore, 2011). According to Immerzeel et al. (2011), the situation will worsen. Based on climate change projections, two-thirds of the region's economies will have less than $200 \, m^3$ per capita per year by 2040–2050. Unmet water demand for the whole MENA region will also increase from the current 16% to 37% in 2020–2030 and 51% in 2040–2050.

Four factors have been identified as the main driving forces of the increase in freshwater demand in the region. The first is the increase in population, which will directly drive an increase in food demands. Population in the MENA region quadrupled during the second half of the twentieth century; it was the second-fastest-growing population in the world until the mid 2000s (Roudi-Fahimi & Mederios Kent, 2007). Recently, total fertility rate has declined to three children in 2006 (from about seven children in 1960) in Lebanon, Egypt, Iran and Tunisia (Haub, 2006). Population growth in the region has slowed, and it is expected to level out after 2050 (United Nations, Department of Economic and Social Affairs, Population Division, 2013). A second driving force is the additional water demands associated with rising standards of living, industrial activity and energy demands, which find their impulse from the socio-economic development in the area. For example, urbanization will impact future water choices in MENA, as most of the population growth has occurred in urban areas (Tropp & Jagerskog, 2006). A third factor, related to the rising living standards, is the dietary shift towards a higher calorific intake of (water-intensive) animal-based products. These impacts will be determined by meat production systems, which are very heterogeneous both in farm practice and in geography (Steinfeld, Mooney, Schneider, & Neville, 2010; Thornton & Herrero, 2010), and whose sustainability varies greatly with the production system (Ridoutt, Sanguansri, Nolan, & Marks, 2012). The fourth factor driving a future increase of water demand in MENA is the expected impact of climate change on the region's water resources and agricultural land, which will affect not only the agricultural sector (with changes in precipitation, temperature, and evapotranspiration rates) but also the function and operation of existing infrastructure, such as hydropower, drainage and irrigation systems (Intergovernmental Panel on Climate Change, 2008). All these factors are driving an increase in water resource demands. Present (and future) challenges mainly regard water availability for food production due to the large amount of water necessary for agriculture.

While these forces drive an increased water demand, which may exacerbate MENA's physical water resource scarcity, the reasons for water problems lie in the capacity of agricultural sectors, local governments and international institutions to respond and adapt to the region's resource scarcity (Allan, 2001). The water-scarcity predicament of MENA has been compounded by large-scale water management problems, such as over-exploitation of aquifers, deteriorating water quality, rationed water supply and suboptimal irrigation services (World Bank, 2007). These problems also have negative impacts on human health, on agricultural yields and on water ecosystem services (Immerzeel et al., 2011).

Agriculture is the largest consumptive water user in MENA. Water consumption by households and industry in the region is relatively small compared with allocations to irrigated agriculture, despite the poor economic returns (per unit volume of water) of irrigated farming (Richards & Waterbury, 2008). Agricultural water withdrawal is over 85% of total water withdrawals in many of the region's economies (Food and Agriculture Organization of the United Nations, 2014) and is mainly allocated to irrigated cereals. The highest irrigation potential in the region can be found in Turkey, Iran and Iraq (Food and Agriculture Organization of the United Nations, 2013). Irrigation systems in MENA are mainly dependent on groundwater resources (Food and Agriculture Organization of the United Nations, 2011). Therefore, declining aquifer levels and extraction of non-renewable groundwater present a growing risk to food production systems in the region. The predominant source of water in the region, however, is soil water. The percentage of rainfed agriculture in the region's farmed lands stands at 75% (Food and Agriculture Organization of the United Nations, 2014).

Current and future food production in MENA is limited not only by the availability of water but also by the extension of arable land. Some of the MENA economies – such as Qatar, United Arab Emirates, Bahrain and Kuwait – have the lowest area of arable land per capita in the world (Food and Agriculture Organization of the United Nations, 2013). With these limitations, most MENA countries became large net importers of agricultural products and especially food commodities, such as cereals, sugar and cooking oils (World Bank, 2003; Ahmed, Hamrick, Guinn, Abdulsamad, & Gereffi, 2013). International trade of agricultural commodities has played a key role in meeting the food requirements of the MENA populations, while complying with the limited water resources in the area (World Bank, 2007). This implicit trade in water has provided the MENA countries with food-water security for the past few decades (Allan, 2001; 2002) and has been enabled by a long-run downward trend in global prices of food commodities (Rosegrant, Paisner, Meijer, & Witcover, 2001). However, it has also enabled policy makers in the region to avoid the implementation of reforms necessary to improve water resources management (Allan, 2001; Zeitoun, Allan, & Mohieldeen, 2010). Promoting economic diversification and growth in the MENA economies is fundamental in order to manage water sustainably – financially, socially and politically (World Bank, 2007). The extent to which the MENA economies rely on trade to secure their food and water needs will be the focus of the following sections.

Background concepts

This section presents the concepts of 'virtual water' and 'virtual water trade' deployed in the analysis of MENA water and food security. The characteristics, opportunity costs and externalities associated with the use of the different types of agricultural water are also discussed.

Virtual water and virtual water trade

The term 'virtual water' describes the water needed to produce a commodity. It can be understood as the water that is 'embedded' within the commodity as a factor of production. Consistently with international trade theory, the trade of commodities can be seen as an implicit exchange in the factors of production, therefore resulting in a trade of virtual water, as first identified by Tony Allan (1993). Virtual water trade can analytically capture the close relationship between water, food security and trade, through the mechanism of redistribution of water resources at a global level. International food trade, and the associated virtual water trade, in fact play a key role in supplying an otherwise limiting resource for the population (see e.g. D'Odorico & Rulli, 2013). Assessments of virtual water trade allow us to quantify the extent to which water-scarce countries rely on water resources available abroad to secure their needs as well as to highlight possible global water savings associated with international trade (see e.g. Chapagain, Hoekstra, & Savenije, 2006). Virtual water trade also enables the analysis of global and national virtual water balances (see e.g. Hoekstra & Mekonnen, 2012; Tamea et al., 2013) and, although the concept has been questioned for not being (by itself) an effective tool for policy development (Wichelns, 2010a, 2010b, 2011), it provides a useful analytical concept to investigate the nexus between water, food and trade. The terms 'trade', 'flows', 'imports' and 'exports', when associated with virtual water, will be used without quotation marks although it is recognized that it is goods that are being traded, not water (Merett, 2003).

The term 'food-water' will also be used to refer to water embedded in food products, in contrast to 'non-food water', which is the water consumed by industry and households (Allan, 2013a, 2013b). Food-water accounts for the overwhelming majority (about 90%) of the water used consumptively by individuals, while non-food water accounts for the small remaining percentage of an individual's water needs (Allan, 2013a; 2013b). Most of the world's economies are food-water insecure, and the MENA economies are in this category.

Green and blue water resources

The present study distinguishes between the two sources of food-water, 'blue' and 'green' water resources. Green water is water that, originating from precipitation, stays in the soil of the root zone to support plant growth. This source of water was identified by Falkenmark (1995) as in contrast to 'blue water', i.e. water stored or flowing in surface and groundwater bodies (Falkenmark & Rockström, 2004). Green and blue water are both involved in food production. The former sustains global rainfed agriculture as well as ecosystems and ecosystem services related to terrestrial systems (Barron et al. 2012). The latter is diverted from surface bodies or withdrawn from aquifers to irrigate crops or to enter the life cycle of commodities through food processing and product transformation, while also being fundamental for ecosystems and ecosystem services related to aquatic systems (Barron et al. 2012). Originating from rainfall, green water is commonly regarded as "a 'free good' in terms of supply" (Yang & Zehnder, 2008, p. 9). It is accessible only to plants and cannot be directly manipulated by human management. Water resource planning and management have traditionally focused on blue water, although it represents only one-third of the world's freshwater resources (Falkenmark & Rockström, 2006). Rainfed agriculture is in fact the world's predominant agricultural production system, thus calling for greater attention to the green water resources of the major food-producing countries (Rosegrant, Cai, & Cline, 2002; Food and Agriculture Organization of the United Nations, 2011; Food and Agriculture Organization of the United Nations, 2013).

Distinguishing the two sources of food water is particularly relevant for promoting more sustainable food production (Gilmont & Antonelli, 2013). Green and blue water differ in terms of alternative uses and opportunity cost, i.e. the value of the alternative that is forgone whenever a choice (between different uses) is made. Agriculture and ecosystem services are the only two competing uses for green water. Compared to blue water, the opportunity cost of green water use is far lower, as it cannot be reallocated to other uses (Hoekstra & Chapagain, 2008; Yang & Zehnder, 2008). As green water use for vegetation growth generally yields lower economic value than crop production, using green water for agricultural production is generally efficient in terms of opportunity cost (Yang, Wang, Abbaspour, & Zehnder, 2006). Blue water, by contrast, is the resource with the highest economic potential, as it can fulfil municipal, industrial and service functions, besides supporting vegetation in conjunction with green water. Blue water in irrigated agriculture yields the lowest economic value among all other use options (Zehnder, Yang, & Schertenleib, 2003). As Allan (2012, p. 6) argued, "the decision to irrigate always leads to a form of blue water over-allocation". For these reasons, it has been argued that, as green water has a lower opportunity cost than irrigation water, trading green water through virtual water trade is overall more efficient than trading blue virtual water, holding other factors constant (Yang & Zehnder, 2008). Moreover, green water use in agriculture is associated with relatively few negative environmental externalities, i.e. the costs (or benefits) that affect other parties without being reflected in the cost of water use (spill-over effects). However, both green water use and blue water use in agriculture result in degradation of water quality through the use of fertilizers and pesticides (Aldaya, Allan, & Hoekstra, 2010). Green water has so far been largely neglected by policy makers and it has not been taken into account in the analyses of national water budgets of the MENA political economies. Nevertheless, it is the main source of agricultural water in the region, underpinning long-established but suboptimal dryland winter farming and grazing (Chatterton & Chatterton, 1996; Allan, 2001; Food and Agriculture Organization of the United Nations, 2014). It is important to distinguish the various water components in policy formulation, as they have different characteristics and opportunity costs, and also in order to appraise the gains and losses of international trade in virtual water.

Methodology

The patterns of virtual water trade in agricultural products were reconstructed by considering 309 products which were traded internationally in the period 1986–2010, as reported by the Statistics Division of the Food and Agricultural Organization of the United Nations (FAOSTAT). These products include crops, animal products, luxury foods (lux-foods), and non-edible agricultural commodities, the latter being considered even if not strictly included in the 'food-water' definition. The reason is that, on the one hand, the water embedded in non-edible crops accounts for a small fraction (10%) of the water traded as embedded in food commodities globally (Carr et al., 2013); on the other hand, non-edible crops contribute to the region's economy and impose pressure on the water resources for other food crops. The full list of 309 commodities considered here can be found in Tamea et al. (2013, Supplementary Material, Table A).

Yearly trade data from FAOSTAT were converted into virtual water flows by considering the virtual water content of each commodity as provided by the global assessment of Mekonnen & Hoekstra (2010a; 2010b). Conversions were differentiated among the countries, assuming that commodities are produced in the countries of origin of the flows. The assessment of virtual water content by Mekonnen and Hoekstra (2010a; 2010b) is based on a distributed model of soil water balance forced by hydrological inputs, which expresses evapotranspiration from cultivated areas and the associated agricultural yields. Country-specific estimates result from the spatial average over the cultivated areas of the country. Given the temporal variability of hydrological forcings and agricultural factors, virtual water content is provided as the average over a period of 10 years (1996–2005) to level out interannual fluctuations and produce robust estimates. Thanks to the centrality of the averaged period with respect to the time span of the present analysis (1986–2010), these estimates of the virtual water content of commodities are taken as reliable throughout the whole span, similarly to, among others, Hoekstra and Mekonnen (2012). Although these estimates do not take into account temporal variations due to changes in efficiency, technology improvements or climate variability, such variations are indirectly expressed by the interannual change of traded goods, which reflect also changes in agricultural production (e.g. reductions during droughts and adverse events or increments due to agricultural expansion and technological advances).

Further details on the assessment of virtual water trade used in the present work can be found in Tamea et al. (2013) and Carr et al. (2013). With respect to such studies, two main methodological innovations are introduced here. First, the trade data underpinning the reconstruction of the patterns of virtual water trade are those reported by the world's exporting economies rather than other approaches deployed in previous studies, as these data are found to be more reliable in the closure of national virtual water balances.

Secondly, the present study differentiates the two sources of virtual water embedded in global virtual water trade, namely green and blue water, whereas the mentioned studies considered only the sum of the two (i.e. total virtual water flows). The proportions of green and blue virtual water embedded in traded commodities have been reconstructed on the basis of the green and blue water footprint estimates provided by Mekonnen and Hoekstra (2010a; 2010b).

Virtual water trade in the MENA region

The MENA region, due to the combination of socio-economical factors and adverse hydro-climatic conditions, largely relies on imports of agricultural products (food in particular) and therefore virtual water. What follows is an extensive quantification and characterization of the virtual water trade implicit in international trade of agricultural products related to the region as a whole and to the single countries in the area. The MENA region considered in this analysis includes the region's developing economies, as classified by the World Bank (2013), and some higher-income water-scarce countries, namely Israel, the Gulf Cooperation Council countries, and Turkey.

MENA in the wider context

At a first instance, the MENA countries are compared to other areas in the world in terms of per capita import of virtual water. The sums of all virtual water volumes imported by the countries in each area, divided by the total population in the area, are shown in Figure 1. The MENA countries are second only to the European countries, whose extremely large per capita virtual water imports (more than double those of other areas in 1998) can be explained by the substantial integration of the region's countries and the proliferation of

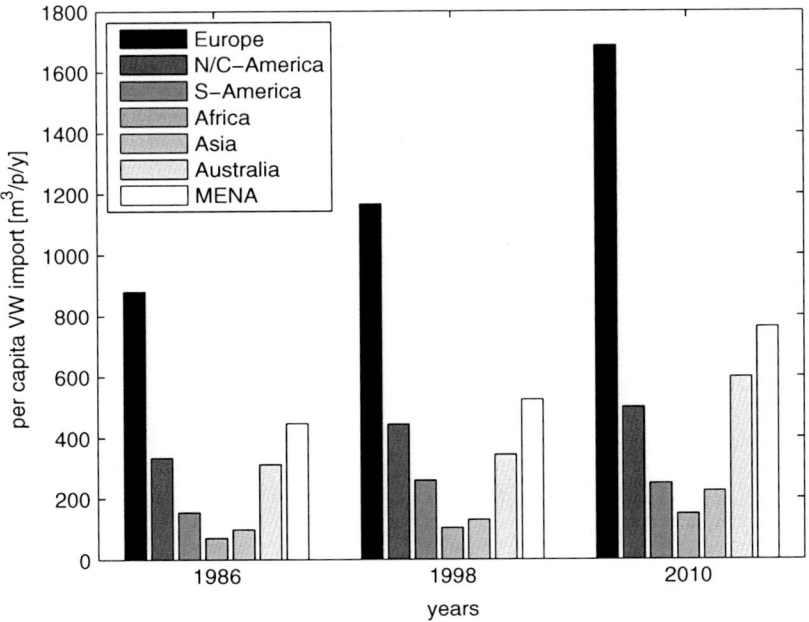

Figure 1.   Virtual water imported by countries in different regions of the world in 1986, 1998 and 2010, in m³ per person per year.

trade agreements that have spurred an intensification of the exchange of goods in the area. Lower amounts (but close to MENA's) are observed in North/Central America and Australia, followed by South America. The overall temporal trend of per capita virtual water import is markedly increasing, dominated by the evolution of European imports' reaching almost 1700 m^3 per person per year in 2010. MENA countries are characterized by a rise from 378 m^3 per person per year in 1986 to 601 m^3 per person per year in 2010. In other areas we see a burst in Australia and a slow-down in the Americas in 2010, whereas virtual water imports in other countries grew steadily.

Total MENA virtual water trade

The total volume of virtual water traded by all the MENA countries (Table A1 in the online supplemental data, available at http://dx.doi.org/10.1080/07900627.2015.1030496) is calculated as the sum of the virtual water volumes imported and exported by each country in the area, differentiated into green and blue virtual water. Virtual water internally exchanged within the region is also quantified and corresponds to the portion of total MENA exports directed towards other MENA countries (or the portion of total MENA imports originating in the MENA area). All volumes are computed as a 25-year average, 1986–2010, including the zero-values occasionally occurring in trade data as a result of low or intermittent exports or occasionally missing (unreported) data (e.g. Iraq, or

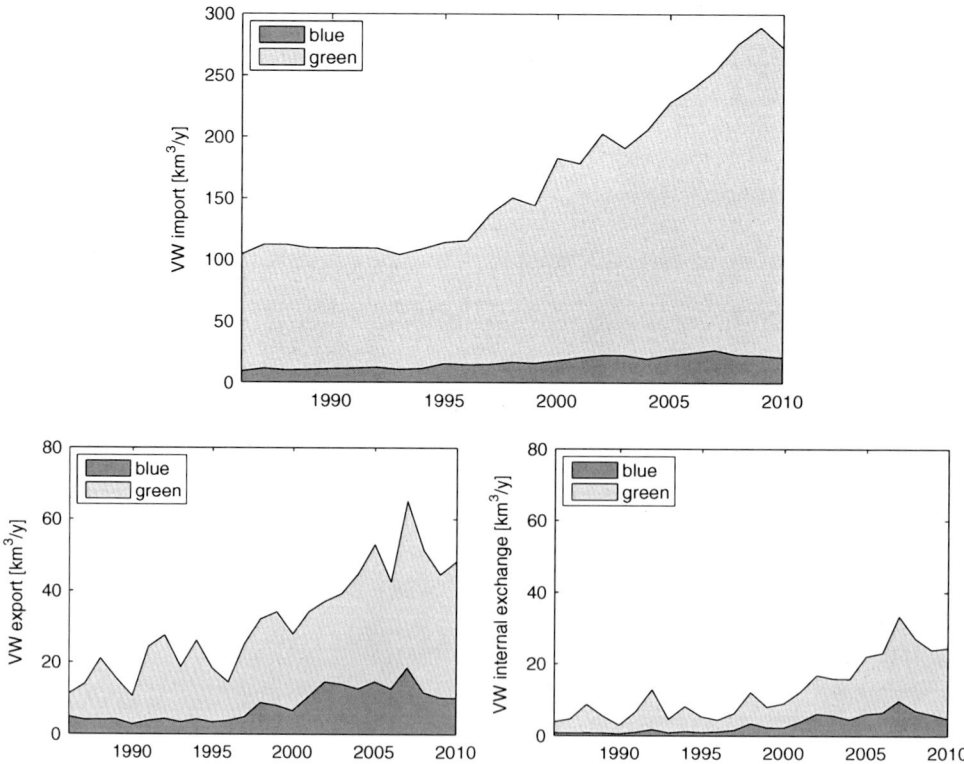

Figure 2. Temporal evolution of green, blue and total virtual water volumes imported (top), exported (bottom left) and exchanged internally (bottom right) by MENA, in km^3 per year.

Egypt in 2010). The temporal evolution of virtual water trade in the aggregated MENA countries is shown in Figure 2.

The data highlight the large volumes of total virtual water traded by the MENA countries. Imports reached 273 km^3/y and exports 480 km^3/y in 2010, twice and four times, respectively, the volume in 1986. Total imports increased substantially over time, with greater contributions from Egypt, Turkey and the United Arab Emirates, followed by Saudi Arabia and Iran. The increased globalization of virtual water trade was facilitated by a downward trend in food commodity prices, which fell by 35% globally between 1980 and 2001. Declining food prices operated to the advantage of the world's food-importing economies, including those of MENA (Allan, 2001). The 2007–2008 price spikes seem not to affect virtual water imports, which keep on increasing steadily until 2009, although the price of food was 50% higher than in 1980 and twice as high as in 2000 (Marktanner, Das, & Niazi, 2011). The inelasticity of the demand for the imported goods can be explained by the fact that staple cereals account for the largest share of food (and virtual water) imports of the MENA countries. In 2007–2008, the price for oil and fats increased by 106%, cereals 83%, dairy products 56%, sugar 26% and meat 22%. These shocks increased MENA's spending on food imports and also spurred domestic food inflation, especially in Qatar, Jordan, United Arab Emirates, Oman, Egypt, Lebanon, Kuwait and Libya (Al Masah Capital Limited, 2012). Even when international food inflation turned negative, domestic food inflation in MENA remained positive. Consumer price inflation remained relatively higher in those MENA countries which are small producers or importers of oil and gas, including Egypt, Jordan, Lebanon, Morocco, Tunisia and the Palestinian Authority (O'Sullivan, Rey, & Galvez Mendez, 2011). Such countries depend on food imports to meet their needs for food and fuel, and they lack the financial resources to implement compensatory measures to dampen price volatility. Inflation remained lower, instead, in countries rich in oil and gas, such as Algeria, Iraq, Syria and Yemen, as well as the Gulf Cooperation Countries and Libya, affecting the low and middle-income segments of the MENA society (O'Sullivan et al., 2011). It has been argued that high and volatile international food prices, which in turn spurred an increase in domestic food prices, were among the triggers of the Arab Spring protests (Breisinger et al., 2010; Woertz, 2011).

According to Figure 2, total virtual water exports from MENA over the whole period were far smaller than virtual water imports. Virtual water exports show several fluctuations that reflect the agricultural production volatility in the region, which have been shown to be caused by the climatic conditions of the region and the occurrence of severe and periodic droughts (De Pauw, 2000; Shetty, 2006). Fluctuations also reflect the intrinsic variability of small and intermittent export fluxes reported by the MENA countries. Virtual water exports decrease after 2007, following the implementation of government interventions and various economic policies, including export restrictions, in response to the global food-price crisis (Breisinger et al., 2010).

Green and blue virtual water trade in MENA

While depicting the temporal evolution of MENA virtual water trade, Figure 2 also provides differentiation into the green and blue water components on which this study is focused. Around 1996, the total virtual water imports started rising, after a steady period. The largest share of virtual water import was green, which is consistent with other studies (e.g. Fader et al., 2011) showing that the largest exporters of agricultural products in the world produce mainly under rainfed systems. Average blue water imports to MENA countries have been 16.8 km^3/y, with a marked increase until 2007 and a minor decrease in the following years, during the period of protests. Blue water import has been around 10% of total imports on average, with a slight decrease in the last decade. The increase in virtual water imports was accompanied by a dramatic decrease in the region's per capita renewable blue water availability. It fell by about half, from 2000 m^3 per person per year in 1986 to 1200 m^3 per person per year in 2010 (Food and Agriculture Organization of the United Nations, 2014). These values, however, do not consider green water, which, if added, would substantially increase water availability in a number of MENA economies (Gerten et al., 2011).

In 1986 blue water accounted for 43% of total virtual water exports and fell to 21% in 2010. This is to say that blue water exports increased over the years (from 4.9 km^3/y to 10 km^3/y), but at a slower pace than total virtual water exports. After an initial steady decade, lasting until 1996, blue water exports increased significantly, thanks to the increase in exports supported by the continuous development of irrigation in the region; whereas it decreased in most recent years, settling to doubled values over the 25-year period. Iran, Egypt and Syria had a very large increase in blue water exports, whereas other countries show a more stable growth.

Virtual water exports have a major intra-regional component because about 25% of total MENA exports are directed towards countries in the same area. The fraction of total imports to MENA countries originating inside the MENA area, on the other hand, is very low (about 7%), indicating that major water transfers occur from other areas of the world. The fraction of blue water sent outside the MENA area was greater than the fraction remaining within the area until 2005; since then it has been smaller.

Major partners of virtual water imports of MENA

The origins of the virtual water imports of the aggregated MENA countries, averaged over the 25-year period considered, are indicated in Figure 3. Ranking the origins of virtual water flows based on average fluxes may not be representative in the case of large inter-annual fluctuations or discontinued trading links (for example in the case of the Russian Federation, which came into action only after 1991), but it enables us to show major on-average flows with single trade partners. The countries of origin of the green and blue virtual water inflows are shown in two different graphs, where the bars indicate the ranked aggregated virtual water flow from each country and the thick increasing curve shows the cumulative percentage of total virtual water import of the aggregated MENA countries. Figure 3 shows that the United States gives by far the greatest contribution to the MENA import, in terms of both green and blue virtual water. Green virtual water inflows from the United States reach 23 km^3/y, followed by Argentina, Australia and Brazil (about 12–13 km^3/y each). These four countries sum up to 40% of the total green virtual water imported by MENA. Besides the United States, blue virtual water inflows originate from Pakistan, France, India and Spain (1 to 2.7 km^3/y). Critical is the situation of Pakistan which, although suffering from internal water deficit (a total water footprint of national consumption greater than the water footprint of production), is still a net exporter of virtual water – especially blue water (data from Mekonnen & Hoekstra, 2011). The United States and India are net exporters of virtual water. Some river basins in these areas suffer from blue water scarcity (Hoekstra & Mekonnen, 2011).

It is worth noting the positions of the MENA countries in the ranked origins of virtual water flows. The MENA country providing the largest volumes of green water to the rest of the area is Turkey (12th place among all sources), with an outflow towards the region of

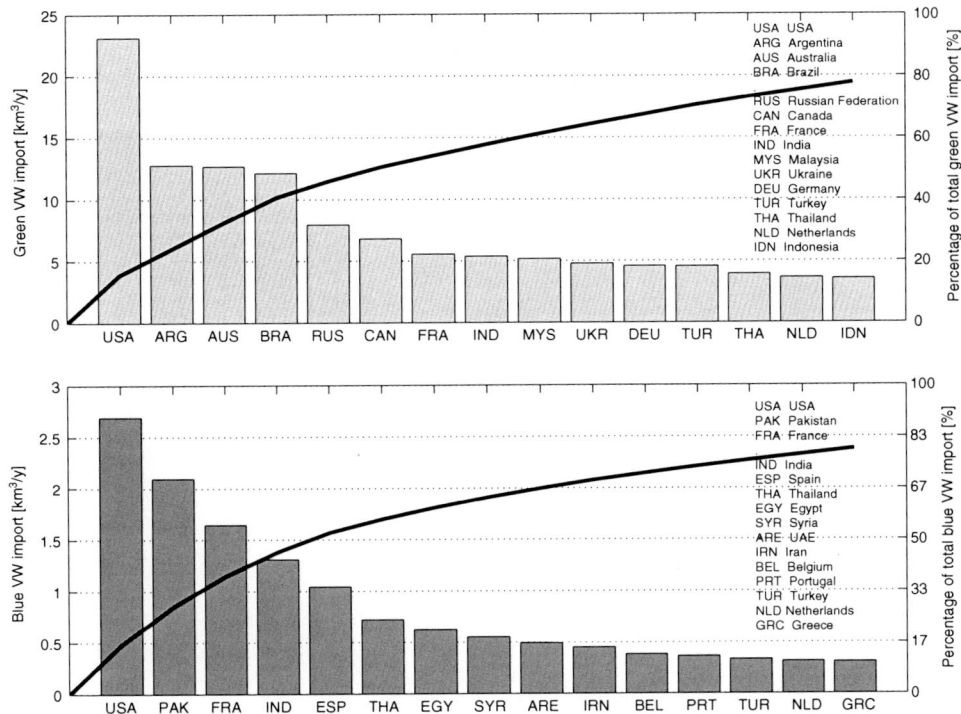

Figure 3. Major trade partners of the MENA area for green (above) and blue (below) virtual water import: average fluxes in the period 1986–2010, in km³ per year. The thick increasing line indicates the cumulative percentage of virtual water imports (axis on the right hand side).

about 4.5 km³/y. When considering blue virtual water flows, a number of MENA countries enter the ranking. Egypt (7th place at 0.65 km³/y), Syria, United Arab Emirates, Iran and Turkey (13th place at 0.35 km³/y) are major blue virtual water import trade partners of the aggregated MENA area. According to data from Mekonnen and Hoekstra (2011), Egypt, Syria and Iran show an internal water deficit. United Arab Emirates compensate for this water deficit through virtual water imports, whereas Turkey shows a positive virtual water balance, with a water footprint of agricultural production that is larger than the water footprint of national consumption (Mekonnen & Hoekstra, 2011).

Major partners of virtual water imports of MENA by category of products

The composition of virtual water imports is analyzed by distinguishing four categories of agricultural products: crops (cereals, fruits and vegetables and the results of their processing); animal products (live animals, primary and derived products of animal origin); lux-foods (coffee, cocoa, sugar, alcoholic beverages and spices); and non-edibles (plant fibers, oil cakes and animal hides). The composition of each category is detailed in Tamea et al. (2013).

Figure A1 (in the online supplementary data) shows, for each category, the five major virtual water import trade partners of the MENA region, distinguishing between green and blue virtual water flows. The proportion of virtual water import associated with crops is by far the largest (119.1 km³/y), followed by lux-foods (29.1 km³/y), animal products (13.6 km³/y) and non-edibles (4.6 km³/y). The fraction of import associated with each category mimics the composition of global virtual water trade (see Carr et al., 2013, for more details), although the crop fraction in MENA import is larger than in global trade (72% versus 54%) and the lux-foods fraction is smaller than in global trade (17% versus 27%). This indicates that MENA countries need, and thus import, more basic food and less high-value agricultural products, compared to the rest of the world. The MENA inflows of virtual water from the United States are mainly associated with the import of wheat, soybeans, maize and cotton lint. Maize and soybeans (including by-products) are also imported from Argentina. Australia and the Russian Federation export wheat and barley to the MENA. The virtual water embedded in the imports from Brazil is mainly associated with cattle meat, maize and sugar; in the case of Australia and New Zealand it is cattle and sheep (live animals, meats and other products). Sugar also comes from France and explains part of the substantial *blue* virtual water fraction associated with the MENA import from this country. The MENA virtual water imports in lux-food also include tea, coffee and cocoa (with their derivatives). Greece and India, besides the United States, provide MENA with virtual water as embedded in their export of cotton lint to the MENA countries.

Country-specific virtual water trade

Virtual water trade of each MENA country

Figure 4 maps the virtual water imports and exports (green and blue) of each country of the MENA area, averaged over 1986–2010, differentiating between green and blue virtual water flows. The maps highlight that Egypt is the country with the largest imports of green water, followed by Iran, Saudi Arabia and Algeria. Large blue virtual water imports are observed in Turkey, Iran, Saudi Arabia and United Arab Emirates, and to a lesser extent in Egypt and Algeria. Virtual water exports are very large in Turkey (both green and blue water), and the second-most green water exporting country is Tunisia.

The MENA countries that export the largest volumes of blue water are Egypt and Turkey. The former has virtually no green water and thus relies almost exclusively on blue water resources. Turkey has the largest agricultural area in the region and is also the largest

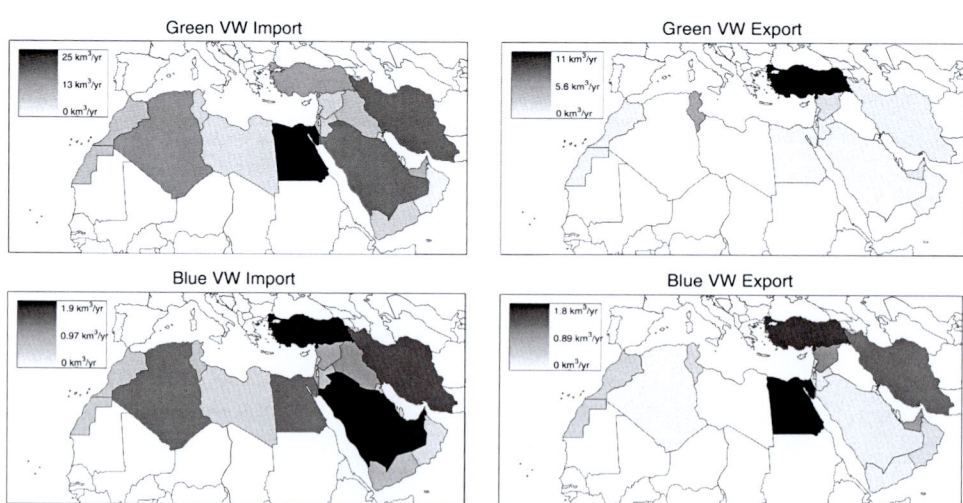

Figure 4. Maps of the virtual water imports (left panels) and exports (right panels) for each MENA country, separated into green (upper panels) and blue (lower panels). Colour saturation is scaled according to average flux over the period 1986–2010, in km^3 per year

producer of agricultural goods (Food and Agriculture Organization of the United Nations, 2014). Other countries with large blue water exports are Iran and Syria, which are characterized by water scarcity and an increasing use of groundwater resources (Food and Agriculture Organization of the United Nations, 2014), which is causing serious problems of water table lowering, groundwater resources over-exploitation and fossil water depletion. Useful information is also provided by a map of the net imports of blue water by MENA countries (Figure A2 in the supplementary online data). This highlights that Egypt and Syria are net exporting countries, while all others import larger (or comparable) volumes of blue water than they export. The blue water balance is particularly negative for Saudi Arabia and to a lesser extent Algeria, whose blue water resources are scarce in comparison to their internal water demand.

Virtual water import dependency of MENA countries

The country-specific analysis can be rounded out by computing a measure of the virtual water dependency of each country on external water resources. The national virtual water import dependency indicator (*D*) is defined as (Hoekstra, Chapagain, Aldaya, & Mekonnen, 2011)

$$D = \frac{I - reE}{WF_c}, \tag{1}$$

that is virtual water imports (*I*) minus re-exports (*reE*) over the water footprint of consumption (WF_c). The latter is calculated from the virtual water balance of each country, that is

$$I + WF_p = E + WF_c, \tag{2}$$

where *E* is the virtual water export and WF_p is the water footprint of agricultural production obtained from converting into virtual water volumes the agricultural production by commodity provided by FAOSTAT, as presented in Tamea et al. (2013).

The FAOSTAT database of commodity trade and the virtual water trade data-set employed in this study do not enable differentiation between exports and re-exports. As it was not possible to quantify the fraction of exported virtual water resulting from the re-export of imported goods, the dependency indicator was built as follows. The range of variability of *D* was identified by considering the two extreme cases: (1) in which there is no re-export of imported commodities (all imported goods are consumed within the country); and (2) in which re-export equals total exports (no internal production is devoted to export). Thanks to the small export fluxes of a number of MENA countries, this range of values is limited and allows taking intermediate values as reference for the following considerations.

The virtual water import dependency is computed for each MENA country in 1986 and 2010, and shown in Figure A3 (in the supplementary online data) as a function of the Human Development Index (HDI) of each country. This index considers three basic dimensions of human development: health, education, and income, measured respectively by life expectancy at birth, mean and expected years of schooling, and gross national income per capita (United Nations Development Programme, 2013). The HDI is here used as an indicator of the socio-economic adaptiveness of the region's economies to analyze whether the dependency on food-water imports is in relation with the diversification and adaptiveness of the MENA countries.

Figure A3 (in the supplementary online data) shows that the MENA countries with a relatively higher HDI generally have a higher dependency on external water resources than countries with lower HDI. The highest dependencies are found in the Gulf Cooperation Countries and Jordan (values above 0.8 in at least one of the two years). Israel also has a substantial dependency on external water resources, especially in the year 2010. The range between minimum and maximum values of D (indicated by the vertical line above and below each marker) is very small, confirming the meaningfulness of the proposed approach. The temporal evolution of virtual water import dependency across MENA has been heterogeneous. The general trend in the region is up, with major increases occurring in Israel, Qatar and Libya (up to 50% more in 2010), and Syria and Morocco (70–80% increase); Turkey's has increased five-fold, mainly because it was very low in 1986. The increase in dependency over time is caused by the marked increase of virtual water imports of such countries, which grew more than proportionally with respect to the water footprint of consumption. Three countries have decreased their virtual water dependency over the period analyzed: Jordan and (to a lesser extent) Algeria and Saudi Arabia. Over the period considered, Saudi Arabia subsidized wheat production through direct and indirect support to farmers, at the expense of local aquifers. In this "unnatural endeavour to make the desert bloom", about 300 km^3 of mostly non-renewable water resources were consumed (Elhadj, 2004, p. 35).

Virtual water import dependency significantly correlates to the HDI in both years (Student's t-test significant at the 5% level). It emerges that dependency on food-water imports is relatively higher in the more diversified and adaptive MENA countries, such as the United Arab Emirates, Kuwait, Bahrain and Qatar, than in the less diversified economies in the region. The most dependent countries are also those that can be considered food secure or moderately food secure, as in the case of Jordan (Breisinger et al., 2010; Woertz, 2011), although high dependency on external water resources implies an exposure to external crises reaching countries through the trade network.

Conclusions

The purpose of the study was to explore the relationship between virtual water, food security and trade in the arid and semi-arid countries of the Middle East and North Africa. The region is extensive and diverse, not only in the scale of the national water deficits but also – and more importantly – in the socio-economic capacity of its economies to cope with the food-related challenges arising from water scarcity. What makes the region's water problems particularly urgent is the combination of a number of socio-economic and environmental challenges that are likely to pose further strain on the already stressed water bodies and that require promotion of more sustainable water management. Since the largest share of societal water consumption is that required for food production, it is of paramount importance to understand how the MENA economies will secure their populations' needs in the next decades. The region's economies are already overwhelmingly dependent on food and virtual water imports. The recent political developments across the region's countries have highlighted the critical role that food security plays in shaping the political and social stability of the area.

In this context, the present study highlighted the role that trade has played over the past 25 years in lessening the water deficits of the MENA economies. The study has explored the water–food–trade nexus using the lens of virtual water in order to account for the water used in the production of food commodities and embedded in the trade between nations. The different sources of water embedded in traded food commodities were also differentiated, revealing that green water is the main source of virtual water imports from non-MENA countries but that blue water accounts for a substantial share of virtual water outflows from the MENA economies. It is important to distinguish green and blue water, as their use is associated with different opportunity costs as well as policy formulation purposes.

Trade in virtual water is much more feasible and affordable than transferring real water resources, and makes it possible for water-scarce countries to effectively cope with poor resource endowments. By analyzing food and virtual water trade in a quantitative and comprehensive way, this study has provided evidence that the political economy of water resources in MENA is dominated by the region's engagement in international trade in food commodities. Given the increasing pressure on local water resources, greater attention should be paid to water resources management in MENA countries but also to trade agreements and exchanges that shift the pressure on water resources outside the area. Although the concept of virtual water has received some critiques from economists questioning its validity as a policy prescription, the role of virtual water trade and virtual water import dependency should be recognized at the policy level, while the increasing pressure on water resources urgently needs to be faced to avoid exacerbation of a problematic socio-economic and environmental situation triggered by food-water insecurity.

Reliance on virtual water imports has enabled policy makers in MENA to minimize the scope of the water challenge and to avoid necessary but politically hazardous reforms in the water sector. It can be argued that economic strengthening of the poorly diversified countries in the region is a prerequisite for developing a virtual water 'strategy' that has the potential to effectively enable them to overcome food insecurity and relieve the pressure on local water resources.

Disclosure statement

No potential conflict of interest was reported by the authors.

Funding

The authors acknowledge funding from the Italian Ministry of Education, Universities and Research [FIR-2012, project RBFR12BA3Y] and the University IUAV of Venice.

Supplemental data

Supplemental data for this article are available at http://dx.doi.org/10.1080/07900627.2015.1030496

References

Ahmed, G., Hamrick, D., Guinn, A., Abdulsamad, A., & Gereffi, G. (2013). *Wheat value chains and food security in the Middle East and North Africa region.* Social Science Research Institute, Duke University. Retrieved from http://www.cggc.duke.edu/pdfs/2013-08-28_CGGC_Report_ Wheat_GVC_and_food_security_in_MENA.pdf

Al Masah Capital Limited. (2012). *MENA food security: Are we doing enough to feed the population?* Dubai, UAE: Al Masah Capital Limited. Retrieved from http://almasahcapital.com

Aldaya, M. M., Allan, J. A., & Hoekstra, A. Y. (2010). Strategic importance of green water in international crop trade. *Ecological Economics, 69,* 887–894. doi:10.1016/j.ecolecon.2009.11.001

Allan, J. A. (1993). *Fortunately there are substitutes for water otherwise our hydro-political futures would be impossible, priorities for water resources allocation and management* (pp. 13–26). London: ODA.

Allan, J. A. (1997). Virtual water: A long term solution for water short Middle Eastern economies? paper presented at the *British Association Festival of Science*, Water and Development Session, University of Leeds, London.

Allan, J. A. (2001). *The Middle East water questions. Hydropolitics and the global economy.* London: IB Tauris.

Allan, J. A. (2002). Water security in the Middle East: The hydro-politics of global solutions. *Columbia International Affairs Online (CIAO).* Retrieved from http://www.isn.ethz.ch/isn/ Digital-Library/Publications/Detail/?ots591=0c54e3b3-1e9c-be1e-2c24-a6a8c7060233& lng=en&id=6839

Allan, J. A. (2012). Food-water security: The role of food supply chains and the agents who operate them. paper presented at the *International Conference on Food Security in Dry Lands, Session: Responsible Investment in Land and Water Resources*, Doha, 14–15 November.

Allan, J. A. (2013a). Food-water security: Beyond water and the water sector. In B. Lankford, K. Bakker, M. Zeitoun, & D. Conway (Eds.), *Water security: Principles, perspectives, practices.* London: Earthscan.

Allan, J. A. (2013b). Acqua e sicurezza alimentare: le filiere dell'acqua alimentare e del cibo. In M. Antonelli (Ed.), *L'acqua che mangiamo.* Isola del Liri, Frosinone: Edizioni Ambiente.

Intergovernmental Panel on Climate Change. (2008). In B. C. Bates, Z. W. Kundzewicz, S. Wu, & J. P. Palutikof (Eds.), *Technical paper VI.* Geneva: Retrieved from http://www.ipcc.ch/ publications_and_data/publications_and_data_technical_papers.shtml#.T7O9RRyR2yY IPCC Secretariat.

Breisinger, C., Rheenen, T., Ringler, C., Nin Pratt, A., Minot, N., Aragon, C., & ... Zhu, T. (2010). Food security and economic development in the Middle East and North Africa. Current state and future perspectives. *IFRI Discussion Paper.* 00985, May, Development Strategy and Governance Division. Retrieved from www.ifri.org

Carr, J. A., D'Odorico, P., Laio, F., Ridolfi, L., & Huerta-Quintanilla, R. (2013). Recent history and geography of virtual water trade. *PLoS ONE, 8*, e55825, doi:10.1371/journal.pone.0055825

Chapagain, A. K., Hoekstra, A. Y., & Savenije, H. H. G. (2006). Water saving through international trade of agricultural products. *Hydrology and Earth System Sciences, 10*, 455–468. doi:10.5194/ hess-10-455-2006

Chatterton, L., & Chatterton, B. (1996). *Sustainable dryland farming. Combining farmer innovation and medic pasture in a Mediterranean climate.* Cambridge University Press: New York.

De Pauw, E. (2000). Drought early warning systems in West Asia and North Africa. In D. A. Wilhite, V. K. Sivakumar, & D. A. Wood (Eds.), *Early warning systems for drought preparedness and drought management.* Proceedings of an Expert Group Meeting, Lisbon, Portugal, September 5–7, World Meteorological Association. Retrieved from www.drought-unl.edu/monitor/EWS/ EWS_WMO.html

D'Odorico, P., & Rulli, M. C. (2013). The fourth food revolution. *Nature Geoscience, 6*, 417–418.

Elhadj, E. (2004). Camels don't fly, deserts don't bloom: And assessment of Saudi Arabia's experiment in desert agriculture. *SOAS Occasional Paper.* School of Oriental and African Studies, University of London. Retrieved from http://www.soas.ac.uk/water/publications/papers/file38391.pdf

Fader, M., Gerten, D., Thammer, M., Heinke, J., Lotze-Campen, H., Lucht, W., & Cramer, W. (2011). Internal and external green-blue agricultural water footprints of nations, and related water and land savings through trade. *Hydrology and Earth System Sciences Discussions, 8*, 483–527. doi:10.5194/hessd-8-483-2011

Falkenmark, M. (1995). Land-water linkages: A synopsis", in "land and water integration and river basin management. *FAO Land and Water Bulletin, 1*, 15–16.

Falkenmark, M., & Rockström, J. (2004). *Balancing water for humans and nature: The new approach in ecohydrology.* London: Earthscan.

Falkenmark, M., & Rockström, J. (2006). The new blue and green water paradigm: Breaking new ground for water resources planning and management. *Journal of Water Resources Planning and Management, 132*, 129–132, May-June10.1061/(ASCE)0733-9496(2006)132:3(129)

Food and Agriculture Organization of the United Nations. (2000). *New dimensions in water security. Water, society and ecosystem services in the 21st century.* Rome: Retrieved from ftp://ftp.fao. org/agl/aglw/docs/misc25.pdf FAO.

Food and Agriculture Organization of the United Nations. (2003). Review of world water resources by country. *Water Reports.* 23. Retrieved from http://www.fao.org/nr/water/aquastat/catalogues/ index2.stm

Food and Agriculture Organization of the United Nations. (2011). *The state of world's land and water resources. Managing systems at risk.* Rome: Retrieved from http://www.fao.org/docrep/017/i1688e/i1688e.pdf FAO.

Food and Agriculture Organization of the United Nations. (2013). *FAO Statistical Yearbook 2013. World food and agriculture.* Rome, Italy: Retrieved from http://www.fao.org/economic/ess/ess-publications/ess-yearbook/en/#.UlXVoRYXJjQ FAO.

Food and Agriculture Organization of the United Nations. on-line database. Retrieved from http://www.fao.org/nr/water/aquastat/main/index.stm [accessed on May 27th, 2014] (2014).

Gerten, D., Heinke, J., Hoff, H., Biemans, H., Fader, M., & Waha, K. (2011). Global water availability and requirements for future food production. *Journal of Hydrometeorology*, *12*, 885–899. doi:10.1175/2011JHM1328.1

Gilmont, M., & Antonelli, M. (2013). Analyse to optimise. Sustainable intensification of agricultural production through investment in integrated land and water management in Africa. In J. A. Allan, M. Keulertz, S. Sojamo, & J. Warner (Eds.), *Handbook of land and water Grabs in Africa. Foreign direct investment and food and water security* (pp. 403–415). Abingdon, UK: IB Tauris.

Gleick, P. H. (2000). *The World's Water 2000–2001: The Biennial report on freshwater resources.* Washington DC: Island Press.

Haub, C. (2006). *World Population Data Sheet.* (Washington, DC: Population Reference Bureau, 2006); and United Nations (UN) Population Division, *World Population Prospects: The 2006 Revision* (2007), data accessed online at www.worldbank.org

Hoekstra, A. Y., & Chapagain, A. K. (2008). *Globalisation of water: Sharing the planet's freshwater resources.* Oxford: Blackwell Publishing.

Hoekstra, A. Y., Chapagain, A. K., Aldaya, M. M., & Mekonnen, M. M. (2011). *The water footprint assessment manual – setting the global standard.* London: Earthscan Ltd.

Hoekstra, A. Y., & Mekonnen, M. M. (2011). Global water scarcity: Monthly blue water footprint compared to blue water availability for the world's major river basins. *Value of Water Research Report Series.* 53, UNESCO-IHE.

Hoekstra, A. Y., & Mekonnen, M. M. (2012). The water footprint of humanity. *Proceedings of the National Academy of Sciences*, *109*, 3232–3237. doi:10.1073/pnas.1109936109

Immerzeel, W., Droogers, P., Terink, W., Hoogeveen, J., Hellegers, P., Bierkens, M., & van Beek, R. (2011). Middle-East and Northern Africa water outlook. World Bank Study. *FutureWater Report*, 98.

Marktanner, M., Das, S. K., & Niazi, A. (2011). Food security challenges in the Arab States/MENA region in the context of climate change: The role of the regional directors team. In *Position paper, Regional United Nations Development Group Arab States/Middle East and North Africa, nexus of climate change and food security.* United Nations Development Group. March.

Mekonnen, M. M., & Hoekstra, A. Y. (2010a). The green, blue and grey water footprint of crops and derived crop products. In *Value of water research report series,* 47. Delft, the Netherlands: UNESCO-IHE.

Mekonnen, M. M., & Hoekstra, A. Y. (2010b). The green, blue and grey water footprint of farm animals and animal products. In *Value of water research report series,* 48. Delft, the Netherlands: UNESCO-IHE.

Mekonnen, M. M., & Hoekstra, A. Y. (2011). *National water footprint accounts: The green, blue and grey water footprint of production and consumption* Value of Water Research Report Series No. 50. Delft, the Netherlands: UNESCO-IHE.

Merett, S. (2003). Virtual water and Occam's razor. *Water International*, *28*, 103–105. doi:10.1080/02508060.2003.9724811

O'Sullivan, A., Rey, M., & Galvez Mendez, J. (2011). Opportunities and challenges in the MENA region. in World Economic Forum, OECD (eds), *Arab World Competitiveness Report 2011–2012*, pp. 42–67.

Richards, A., & Waterbury, J. (2008). *A political economy of the Middle East* (Third edition). Boulder: Westview Press.

Ridoutt, B. G., Sanguansri, P., Nolan, M., & Marks, N. (2012). Meat consumption and water scarcity: Beware of generalizations. *Journal of Cleaner Production*, *28*, 127–133. doi:10.1016/j.jclepro.2011.10.027

Rosegrant, M., Cai, X., & Cline, S. (2002). *World water and food to 2025.* Washington DC: International Food Policy Research Institute [IFRI].

Rosegrant, M. W., Paisner, M. S., Meijer, S., & Witcover, J. (2001). *Global food projections To 2020: Emerging trends and alternative futures*. Washington DC: IFRI.

Roudi-Fahimi, F., & Mederios Kent, M. (2007). Challenges and opportunities the population of the Middle East and North Africa. *Population Bulletin*, 62, 2, Population Reference Bureau, Washington DC.

Sakmar, L. S., Wackernagel, M., Galli, A., & Moore, D. (2011). Sustainable development and environmental challenges in the MENA region: Accounting for the environment in the 21st century. *Working Paper 592*. Retrieved from http://www.footprintnetwork.org/images/uploads/Sakmar_et_al_2011.pdf

Shetty, S. (2006). Water, food security and agricultural policy in the middle East and North Africa region. *Working Paper Series,* 47. Office of the Chief Economist, World Bank. Retrieved from www.worldbank.org

Steinfeld, H., Mooney, H. A., Schneider, F., & Neville, L. E. (Eds.). (2010). *Livestock in a changing landscape*. London: Island Press.

Tamea, S., Allamano, P., Carr, J. A., Claps, P., Laio, F., & Ridolfi, L. (2013). Local and global perspectives on the virtual water trade. *Hydrology and Earth System Sciences, 17*, 1205–1215. doi:10.5194/hess-17-1205-2013

Thornton, P. K., & Herrero, M. (2010). The inter-linkages between rapid growth in Livestock production, climate change, and the impacts on water resources, land use, and deforestation. In *Policy research working paper,* 5178. Washington, DC: The World Bank.

Tropp, H., & Jagerskog, A. (2006). Water scarcity challenges in the Middle East and North Africa. In *Human development report office occasional paper,* 31. SIWI, Stockholm International Water Institute.

United Nations, Department of Economic and Social Affairs, Population Division. (2013). *World Population Prospects: The 2012 Revision*. Volume II, Demographic Profiles (ST/ESA/SER.A/345).

United Nations Development Programme. (2013). *Human development report 2013. The Rise of the south: human progress in a diverse world*. New York: United Nations Development Programme.

Wichelns, D. (2010a). Virtual water and water footprints offer limited insight regarding important policy questions. *International Journal of Water Resources Development, 26*, 639–651. doi:10. 1080/07900627.2010.519494

Wichelns, D. (2010b). Virtual water and water footprints: Policy relevant or simply descriptive? *International Journal of Water Resources Development, 26*, 689–695. doi:10.1080/07900627. 2010.519533

Wichelns, D. (2011). Assessing water footprints will not be helpful in improving water management or ensuring food security. *International Journal of Water Resources Development, 27*, 607–619. doi:10.1080/07900627.2011.597833

Woertz, E. (2011). Arab food, water, and the big land grab that wasn't. *Brown Journal of World Affairs, XVIII*, 119–132.

World Bank. Making the most of scarcity: Accountability for better water management results in the Middle East and North Africa, *MENA Development Report*, The World Bank, Washington D.C. Retrieved from http://siteresources.worldbank.org/INTMNAREGTOPWATRES/Resources/Making_the_Most_of_Scarcity.pdf (2007).

World Bank. (2009). *Water in the Arab World. Management perspectives and innovations*, edited by N. V. Jagannathan, A. S. Mohamed, & A. Kremer. Washington DC: World Bank.

World Bank. (2013). *Online Database*. Website accessed in June 2013.

Yang, H., Wang, L., Abbaspour, K. C., & Zehnder, A. J. B. (2006). Virtual water trade: An assessment of water use efficiency in the international food trade. *Hydrology and Earth System Sciences, 10*, 443–454. doi:10.5194/hess-10-443-2006

Yang, H., & Zehnder, A. (2008). Zaragoza, Spain Globalisation of water resources through virtual water trade. *Proceedings of the Sixth Biennial Rosenberg International Forum on Water Policy*.

Zehnder, A. J. B., Yang, H., & Schertenleib, R. (2003). Water issues: The need for action at different levels. *Aquatic Sciences – Research Across Boundaries, 65*(1), 1–20. doi:10.1007/s000270300000

Zeitoun, M., Allan, J. A., & Mohieldeen, Y. (2010). Virtual water 'flows' of the Nile Basin, 1998–2004: A first approximation and implications for water security. *Global Environmental Change, 20*, 229–242. doi:10.1016/j.gloenvcha.2009.11.003

Rapid assessment of the water–energy–food–climate nexus in six selected basins of North Africa and West Asia undergoing transitions and scarcity threats

Caroline King[a,b,c] and Hadi Jaafar[d]

[a]Centre for the Environment, School of Geography and the Environment, Oxford University, UK; [b]Ecosystems and Human Development Association, Alexandria, Egypt; [c]International Institute for Environment and Development, Edinburgh, UK; [d]Department of Agriculture, Faculty of Agricultural and Food Sciences, American University of Beirut, Lebanon

Existing strategies for management of water scarcity in the Middle East and North Africa negotiate a complex system of trade-offs between water, energy, and food production. The effects of rural households' green water management practices on basin-level water, energy, food and carbon stocks and flows are sketched qualitatively in six basin agro-ecosystems. The case for increased strategic support for green agricultural water management practices appears stronger when weighed from the nexus perspective, rather than purely from the point of view of water balance and food production. Trade-offs under critical transitions affecting agricultural water use are explored, and the scope for quantitative monitoring is discussed.

Introduction

The water–energy–food–climate nexus approach highlights the synergies in human life-support systems between these essential types of ecosystem stocks and flows (Hoff, 2011). In the drylands, a growing population will need to be nourished while balancing competing demands for all sectors of the economy with less dependable water availability in future (Falkenmark & Molden, 2008; IPCC, 2014). Energy use for pumping and treating water can boost water supplies, but increasing emissions further exacerbates climate change and threatens resource-dependent populations (WWDR, 2014). The nexus approach invites water managers to reconsider the inevitability of a bloated blue water and emissions footprint for agriculture by maximizing green agricultural water uses (Herren, 2011; IAASTD, 2009; Keys, Barron, & Lannerstad, 2012) and re-evaluating the integration of water, geochemical and energy cycles in agriculture together with the other sectors (UN, 2014; WEF, 2014).

Implementing the nexus approach requires integrated scientific re-assessment of resource-use efficiencies in relation to the interlinked water, energy and food agendas to better inform national economic development strategies and more sustainable business models (WEF, 2014). Environmental and social impact assessment of existing strategies offers a key tool for retrospective identification and weighing of observed trade-offs. *Ex ante* impact assessment using scenarios and cost–benefit assessment of options enables the exploration of improved strategies for the future to secure win-win trade-offs maximizing benefits across all sectors. Policy design, implementation and monitoring are social learning processes that require a strong experimental design, inclusive stakeholder participation and adaptive management (Gallopín, 2006). Marginalized and low-income populations are critical stakeholders in policy making at the water–energy–food–climate nexus in the drylands due to their resource dependence, knowledge and stewardship practices (Safriel et al., 2005).

A solid consensus on shared objectives and measures of collective progress towards them is critical to unite and motivate all stakeholders in the policy process, including resource-dependent populations. This article explores the application of the water–energy–food–climate nexus frame (Hoff, 2011) as a means to weigh the trade-offs associated with selected green water management practices in six water basin systems across North Africa and West Asia (see Figure A1 in the online supplemental data at http://dx.doi.org/10.1080/07900627.2015.1026436). A three-dimensional hydro-agro-ecological assessment of current effects of rural households' green water management practices on the basin-level water–energy–food balance is presented. Discussion of anticipated future evolution of these balances under ongoing climatic changes introduces a fourth dimension to the assessment. The approach is used to locate and qualitatively weigh the trade-offs between agro-ecosystem inputs and products under different land management practices in search of scope for strategic intervention to increase cross-sectorial efficiencies.

The objectives of the article are:

(1) to broaden decision frames from available basin-level water-balance analyses of 'blue water' versus the efficient 'green' use of soil moisture to consider stocks and flows of geochemicals and energy
(2) to uncover and characterize the synergistic balances of energy and nutrients inherent in the available water-balance scenarios with and without green water management practices from localized action research findings and available decision-support tools applied in six water-scarce basins
(3) to make recommendations for adaptive management strategies to monitor, adapt and maximize synergies using measurements to assign values to the units of water, energy and food stocks and flows.

Assessing the benefits of green agricultural water management practices in agro-ecosystems

Blue water flow is the total runoff, including the sum of surface runoff (produced from the partitioning of rainfall at the land surface) and groundwater recharge (produced from the partitioning of soil water in the soil profile) (Rockström, 1999). Green water is the return flow of water to the atmosphere as evapotranspiration, which includes a productive part as transpiration and a non-productive part as direct evaporation from the soil, lakes, and ponded areas, and from water intercepted by canopy surfaces. Farmers' green water management practices focus on ensuring that there is sufficient water within the soil profile to support crop productivity, adding irrigation where needed to supplement soil water (IAASTD, 2009).

Although resource users' primary objective in green water management is usually the maximization of crop productivity and income, it has been observed that other ecosystem benefits may also result from conserving evapotranspiration, net primary production and resilience in dryland ecosystems (Rockström & Gordon, 2001; Rockström et al., 2010).

Assessments of ecosystem services or multifunctional benefits associated with 'green' or 'sustainable' agriculture have gained increasing attention (IAASTD, 2009; Pretty, 2003; Pretty et al., 2006). In addition to maximizing the availability of rainwater for use, 'green' agriculture usually has low demands for energy and other inputs, and can encourage carbon sequestration (Hoff, 2011). It can also enhance (blue) groundwater recharge (Keys et al., 2012).

The identification of trade-offs within ecosystems requires understanding of the system function, processes and qualities of the stocks and flows of resources that move within them (Ash et al., 2010; Watson & Zakri, 2005). The water–food–energy–climate nexus targets three essential types of resource flows and assets: hydrological, geochemical and energy. At the policy level, through national strategies for water, agriculture and energy, these are usually considered on a sectorial basis, whereas national climate adaptation and mitigation strategies take a cross-sectorial view. Although national strategies do not correspond to the scale of basin systems (which are often either sub-national or international), they are sometimes able to incorporate the findings of basin-level studies (Falkenmark & Molden, 2008), where these are available to decision makers.

Methodology

Six basins were selected for discussion as case studies in this article (Table 1). This selection targets research and pilot-testing sites in rainfed and irrigated agro-ecosystems integrating livestock production and household water supplies maintained by National Agricultural Research and Extension System staff members taking part in the Water and Livelihoods Initiative of the International Center for Agricultural Research in the Dry Areas (ICARDA) (WLI, 2014).

Based on information available as of the end of 2013, a rapid review of rural communities' household budgeting of the selected green water productivity-enhancing practices was made to identify what is known or can be inferred about effects on water availability and consumption, energy and geochemical processes, and food production.

Table 1.   Selected basins in the Middle East and North Africa (partially based on UNDP, 2013).

Agro-ecosystem	Basin	Countries	Basin size $(1000 \, km^2)$
Rainfed with supplemental irrigation from groundwater	Abyan	Yemen	0.536
Rangeland in urbanizing water basin under population pressure and water scarcity	Zeuss Koutine	Tunisia	1.305
Rangeland in urbanizing water basin under population pressure and water scarcity	Jordan River	Lebanon, Syria, Israel, Jordan, Palestine	19.839
Rainfed with supplemental irrigation from groundwater	Orontes	Lebanon, Syria, Turkey	37.900
Irrigated from large river system under increasing water scarcity	Tigris-Euphrates	Iraq, Syria, Turkey, Jordan, Saudi Arabia, Iran	793.314 (combined)
Irrigated from large river system under increasing water scarcity	Nile	11 countries	3173

This review included consideration of previously published research on the water–energy–food–climate nexus in each of the selected basins.

The benefits of resource consumption under different management practices were weighed qualitatively as positive and negative factors. This was based on the use of a simplified resource-accounting approach (UN, 2014) to either quantitatively or qualitatively characterize the opening balance, inflows, storage, products, and outflows of water, energy and geochemicals associated with each land management alternative, based on the information available. Geochemical inputs produce two closing balances: food, and by-products that include waste and pollution hazards.

For presentation and discussion of the case-study results, the available mix of quantitative and qualitative information on water, energy and geochemical stocks and flows in the selected basins is tabulated in a binary representation. This approach is similar to those used in the future scenarios of the millennium ecosystem assessment (Carpenter, Pingali, Bennett, & Zurek, 2005) and popular assessments of sustainable land management strategies (WOCAT, 2013). This presentation provides a simple indication of the direction of change in each closing balance of each natural capital stock as either positive or negative.

Rapid nexus assessment of agricultural water management practices in six catchments in transition

Transitions from rainfed to groundwater-irrigated agro-ecosystems: horticultural production in the Abyan Delta and the Orontes Basin

In the Abyan Delta and the Orontes Basin, as in much of the region, a transition has taken place from reliance on periodic rainfall events to use of groundwater for irrigation. This reduces present climate-driven uncertainty, but increases energy use, emissions and future water scarcity by lowering the water table. Increased certainty enables higher input agriculture, including horticultural production. This requires and rewards the increased use of agrochemicals, which can affect the environment. Export horticulture also requires long-distance transportation and other water- and energy-intensive processing.

The Abyan Delta basin (Figure A2 in the supplemental online data) covers 53,000 ha. It receives less than 100 mm average annual rainfall in the lower reaches but often considerably more in the upper reaches. Irrigation is traditionally achieved through a system whereby seasonal floodwater travelling through usually dry ephemeral river systems (known as *wadis*) is conveyed to irrigable fields. This is known as spate irrigation because the floodwater flow pattern is characterized by discrete events, which last for only a few hours. These create water discharges and recession flows, usually lasting only one to a few days. The flow is diverted from the *wadi* channels into typically short steep canals that convey the water to bunded basins, which can be inundated to depths of 0.2–2 m (Mehari, Van Steenbergen, & Schultz, 2011).

In Abyan, horticultural production has expanded in recent years. This is generating high-value crops, and can therefore be considered by the farmers who are growing them to have a positive effect on productivity (Table 2). Horticultural crops such as bananas require continuously available water to be pumped from the ground, as well as inputs of agrochemicals, which are believed to have affected biodiversity and reduced honey production in the delta (ICARDA, 2013). The human time and energy required for management of the spate irrigation system is reduced when groundwater pumping systems are installed. But the use of fuel for energy and the resulting carbon emissions are increased. Rich farmers who can dig wells and invest in inputs benefit from the transition,

Table 2.   Transition from spate water to groundwater use in the Abyan Delta basin agro-ecosystem.

	Balance of ecosystem stocks and flows
Water	↓
Food and income	↑
Natural cycling and storage of geochemicals	↓
Carbon sequestration and emissions reduction	↓

but the free supply of water and good soil conditions that the spate systems used to provide to poor farmers are degraded.

The total renewable annual water supply to the spate irrigation system through the four main *wadis* in the Abyan Delta has been affected by changes in the climate, in addition to the anthropogenic changes associated with the expansion of horticultural production (Atroosh & Moustafa, 2012; RoY, 2013). The total volume of water available in the Abyan Delta basin has been calculated at 156 MCM (million cubic metres) per year. This is supplemented by increasing volumes of groundwater use, currently estimated at 112 MCM/y, of which only 84 MCM/y is recharged (Atroosh, 2013). Although there is scope for improvement of the water accounting, it is already evident that the groundwater table is falling and that saline intrusion is occurring in coastal areas (UNDP, 2013).

The transition from spate water to groundwater use may have a positive effect on food and income production for some farmers, but the negative effects on water availability in the spate system, the underlying aquifer and the soil profile will have negative effects on others. The natural cycling and storage of geochemicals, carbon sequestration and emissions reductions are also negatively affected by the altered cultivation pattern (Table 2).

Good agricultural practices, including the use of agroforestry to create favourable microclimates, selection of drought-resistant varieties, maintenance of local biodiversity, integrated pest management, reduced application of agrochemicals, and water harvesting can save on input costs, reduce environmental pollution, improve soil quality and sequester carbon (IAASTD, 2009; Massaad & Jomaa, 2013; MoE, 2012). As illustrated above, large-scale horticulture production, particularly for export, can discourage these traditional green water management practices developed by generations of land and water users in the drylands. However, market incentives can also reward and reinforce them, e.g. by enabling access to organic and other certification- and labelling-dependent markets (WLI, 2014).

The total area of the Orontes Basin is 37,900 km², including parts of Lebanon, Syria and Turkey (Figure A3 in the supplemental online data). Lebanon's Beka'aa Valley covers 17,000 ha at the upstream end of the basin, and receives average annual rainfall of 150 mm (total volume: 25 MCM/y). In the Beka'aa, imported commercial varieties of grapes (Superior table grape), and soft fruits, are increasingly cultivated in expanding orchard plantations, using groundwater for irrigation.

Changes in land and water use in the Beka'aa have been captured through studies of the variations in summer agricultural vegetative cover over the last decade (Jaafar, 2015). Overall, the agricultural land cover visible in summertime has expanded by more than 20% on average since the beginning of the century, mainly in areas not served by the existing open-channel irrigation scheme. The water required for these lands emanates from the groundwater of the Orontes Basin.

Retaining rainfed cultivation of the drought-resistant local fruit-tree varieties, harvesting rainwater in *negarims* around the bases of the trees and using integrated pest management practices increases resilience to drought, pests and disease, conserves a diversity of fruit tastes, and maximizes the productive use of available rainwater (IAASTD, 2009; Massaad & Jomaa, 2013). Farmers' use of good practices can lead to beneficial effects on stocks and flows of water and food, natural cycling and storage of geochemicals, carbon sequestration and emissions reductions (Table 3). However, it is difficult to assess whether these are enough to counter the effects of the increasing use of energy to pump groundwater for irrigation over the increasing cultivated land area.

The second national communication of Lebanon to the United Nations Framework Convention on Climate Change (UNFCCC) identifies emissions from agriculture, including those associated with enteric fermentation, manure management, soils and field burning of crop wastes, but not water pumping or treatment (MoE, 2011). The water resources and irrigation sector does not have a separate emissions inventory. Vulnerability and adaptation assessments are presented for the water sector and the agricultural sector, but not for the energy sector. It is not possible to know the volume of emissions, nor how much water is being pumped, nor what volumes of agricultural chemicals are being applied.

The total volume of flow through the Orontes Basin from springs and runoff water amounts to around 400 MCM/y (Comair, McKinney, Scoullos, Flinker, & Espinoza, 2013). International agreements give Lebanon the right to use 80 MCM/y of surface water and 16 MCM/y of groundwater for irrigation in the Beka'aa. This was calculated to allow use of an estimated 7000 m^3 of irrigation water per hectare per year, for 7000 ha. Presently, only 50 MCM/y is believed to be in use to support cultivation on 5000 ha of land, 4600 of which are irrigated (Comair et al., 2013). Springs are recharged annually by the rainwater, including flows through the *wadi* beds, but spring water also includes the natural discharge of stored reserves of palaeowater (Al-Bakri, Suleiman, Abdulla, & Ayad, 2011; Al-Charideh, 2013).

Whether the increase in groundwater extraction will be sustained in the near future has yet to be determined, given unofficial reports of declining groundwater levels in the upstream part of the basin (Jaafar, 2015). Although available studies suggest that there is not an overall deficit in the water balance in the basin, climatic changes are already affecting crop and livestock production in the Beka'aa, and further effects are predicted in future due to both climate change (MoE, 2011) and demographic pressures.

Precision irrigation and water reuse for economic water-use efficiency in the irrigated agro-ecosystems of the Nile Delta and the Mesopotamian Plain

Agro-ecological green water management strategies in irrigated areas of the Nile Delta, Mesopotamian Plain, Abyan Delta, Orontes Basin and across the region continue to focus more tightly on aligning water supply to economically productive crop-water needs and

Table 3. Potential effects of reinforcing good agricultural practices on balance of stocks and flows in horticulture production systems, Beka'aa, Lebanon.

	Without good practices	With good practices
Water	↓	↑
Food and income	↑	↑
Natural cycling and storage of geochemicals	↓	↑
Carbon sequestration and emissions reduction	↓	↑

increasing the use of marginal-quality water in agriculture. Regulation of irrigation volumes and timing to avoid over-irrigation can improve the quality of crop production, including for high-value horticultural crops (FAO, 2012). But reducing water application will also reduce downward percolation to recharge reusable shallow groundwater.

Aligning fertigation and nutrient requirements to plant requirements helps further increase the volume and quality of production achievable per unit of water (FAO, 2012). In systems where water application is reduced, high-quality water is needed, to avoid salinity build-up due to lack of soil flushing. Organic matter renewal, soil water holding capacity and carbon sequestration may also be reduced.

Increasing control of water application through pressurized sprinkler or drip systems requires more energy (El-Qousy, Mohamed, Aboamera, & Kheira, 2006; García, Díaz, Poyato, Montesinos, & Berbel, 2014) (Table 4). Ensuring sufficient water for salt-sensitive horticultural produce such as citrus can involve further energy costs to pump or treat water. Deficit irrigation requires compelling demonstration of plants' responses to water stress and identification of cost savings, improved qualities (e.g. higher sugar in citrus) and increased value of produce to convince and motivate farmers to use these techniques (FAO, 2012).

Many farmers across the region use shallow drainage water and groundwater to reduce their vulnerability to water shortage and increase productivity, because these sources are available at any time on demand, avoiding the uncertainty and variability of irrigation volumes and timing delivered through public or collectively managed systems. These practices have an energy cost, which can range across several orders of magnitude, depending on whether and how the water is treated (UNDP, 2013). Using or reusing saline and marginal-quality water to produce fodder for livestock avoids costs for treatment (although it may still include costs for pumping), adds value to a resource that might otherwise go to waste (Saleh, Ibrahim, Dhehibi, & Hassan, 2013), and enables saving higher-quality water for other uses.

The Euphrates runs through Turkey, Syria and Iraq (Figure A4 in the supplemental online data). In the Iraqi stretch of the river, the Abu Ghraib irrigation system covers around 272,000 ha in the Mesopotamian Plain between the Euphrates and Tigris Rivers. Average annual rainfall is about 123 mm (Al-Falahi, 2014). Due to reductions anticipated in precipitation over the basin (J. P. Evans, 2009), the decrease in annual discharge simulated for the Turkish portion of the Euphrates has been projected to reach 30–70% at the end of the twenty-first century. There is no climate change adaptation strategy for the Euphrates. Iraq has not published a national communication to the UNFCCC, nor any sectorial climate change strategy for water, agriculture or energy. Turkey has prepared strategies to adapt to the anticipated reductions in rainfall and surface water flows in the Turkish portion of the Euphrates (TR, 2011).

Flows into Iraq through the Euphrates have decreased from around 29 km^3 of water per year during 1932–1974 to around 21 km^3/y during 1975–2003 due to upstream

Table 4. Effects of precision irrigation and water reuse on balance of ecosystem stocks and flows in the Nile Delta and the Mesopotamian Plain.

	Precision irrigation	Water reuse
Water	↑	↑
Food and income	↑	↑
Natural cycling and storage of geochemicals	↓ ?	↓ ?
Carbon sequestration and emissions reduction	↓ ?	↓ ?

developments as well as climatic changes (J. P. Evans, 2009). Additional flow from the Tigris is estimated at around $50\,km^3/y$, resulting in a total of $70\,km^3/y$ entering Iraq at the first entry point on the Iraqi–Syrian border. According to the records kept by the Office of Irrigation of the Ministry of Water Resources (MoWR), the total volume of water flowing into the Abu Ghraib canal entry point is 622 MCM/y. The total water requirement for agriculture has been estimated at somewhere close to 700 MCM/y, and the outflow at around 150 MCM/y (Al-Falahi, 2013, 2014).

The sources and volumes of water used in irrigation at Abu Ghraib have been estimated as follows: irrigation network sourced from the Abu Ghraib Canal, 80% (622 MCM/y according to MoWR); groundwater, 17% (132 MCM/y); drainage water, 3% (23 MCM/y) (Al-Falahi, 2013, 2014). Recent studies appear to show a lowering of the groundwater table at Abu Ghraib, suggesting an imbalance between extraction and recharge (Voss et al., 2013). At the same time, salinity problems are exacerbated (Evans, Soppe, Barrett-Lennard, & Saliem, 2012). Although techniques for water saving and reuse offer a means to cope with some of these problems (Hussien, Aoda, & Alfalahi, 2014; Neama, Al-Falahi, & Hamoudi, 2014; WLI, 2014), the implications concerning associated uses of chemicals and energy in agriculture are difficult to assess with any accuracy, and no policy framework is in place through which this could be done systematically.

The Nile Delta extends over around 2,500,000 ha, with maximum annual rainfall of around 200 mm at the coast, and less inland. The cultivated area of the delta is supplemented by ongoing land reclamation. The annual volume of surface water inflow at the High Aswan Dam is $55.5\,km^3/y$, and annual outflow to the sea, $12.2\,km^3/y$ (MWRI, 2010). Present and anticipated future water balance and food production have been intensively studied. Present and potential future uses of precision irrigation and water reuse technologies have also received significant attention (Attaher & Medany, 2008; Nour El-Din, 2013).

Under climate change and upstream development the volume of water reaching the Nile Delta from the Nile Basin (Figure A5 in the supplemental online data) will be reduced. Water demand for drinking in Egypt is $1.8\,km^3/y$, while industry requires 1.4 billion cubic metres. The amount of water used by the agricultural sector in 2010 is about $67\,km^3/y$, including leaching requirements and deep percolation to shallow groundwater. To achieve this, and also support other sectors, other water sources are used besides surface water. These include groundwater and drainage water.

Egypt has prepared two national communications to the UNFCCC (EEAA, 2010b), a National Environmental, Economic and Development Study for Climate Change (EEAA, 2010a), an assessment of potential impacts of climate change on the Egyptian economy and a draft adaptation strategy for the Ministry of Water Resources and Irrigation (Nour El-Din, 2013). The national communication includes assessments of national water resources and the energy and agricultural sectors – these are the two sectors with the highest emissions in the national greenhouse gas emissions inventory.

The water resources and irrigation sector does not have a separate emissions inventory, but basic methods for estimating the energy used to pump the water used in agriculture are available (Attia et al., 2005; Fraenkel, 1986). Vulnerability and adaptation assessments are presented in the national communication for the water and agricultural sectors but not for the energy sector. The balance of effects on stocks of water and food that will be achieved with ongoing technological shifts in irrigation management and water reuse can be expected to be positive. However, there remains a need to better understand the accompanying effects of these transitions on the natural cycling and storage of geochemicals, carbon sequestration and emissions reductions (Table 4).

Increasing urban water and food demands: water harvesting in the Jordan River Basin and the Zeuss Koutine Watershed

Increasing water scarcity, constrained access to pasture resources, decreasing returns from agriculture, and more attractive livelihood opportunities promised by other sectors have led to depopulation of the rural areas, neglect of traditional cultivation practices, increasing reliance on imports of food and animal feed, and out-migration to cities (Sghaier, 2009). As a result, the energy demands of cities to pump and treat water are growing, as are their demands for food imports.

Water harvesting is widely recommended as a solution to support rural communities and sustainable food production in dry areas (Jaafar, 2014; WOCAT, 2013). This practice is recognized to save water, increase productivity, and improve soil and carbon sequestration (Table 5). Where rainwater captured in the soil profile exceeds the water holding capacity, it can percolate downwards to recharge groundwater reserves. Increasing groundwater recharge could help in avoiding the need for the construction of more water treatment plants and slow the increasing energy demands for pumping groundwater from a falling water table in the urban areas. However, because water harvesting is labour-intensive, rural communities often do not anticipate sufficient benefits from investing in it.

Wadi Oum Zessar (surface area: 33,600 ha) is one of several ephemeral rivers feeding the Zeuss Koutine Aquifer in the Koutine Watershed, Tunisia (Figure A6 in the supplemental online data). Traditionally, in the upper reaches of the Koutine Watershed, rainfall and water-harvesting structures (known as *jessour*) support agroforestry, intercropping of barley and natural vegetation for grazing by livestock. Further down, in the plain, water-harvesting structures are known as *tabia*. Harvesting rainwater to support trees and fodder for livestock can simultaneously increase the recharge of groundwater (Ouessar et al., 2009, 2004). Where rainwater captured in the soil profile exceeds the water holding capacity, it will percolate downwards and recharge the groundwater reserves flowing through the Zeuss Koutine Aquifer, serving the downstream urban population in the cities of Medenine, Tataouine, Jarzis, Jerba and Benguerdene.

Maintenance of the water-harvesting systems is labour-intensive and not particularly profitable for private farmers. The Tunisian government has invested directly in increasing recharge to the aquifer through the construction and maintenance of artificial recharge structures in the upper part of the watershed (Hadded et al., 2013). The only direct incentive the rural population has for providing this recharge service is the 10% of groundwater recharge that they extract again at their own cost for local use (WLI, 2013). Economic incentives for smallholders to undertake and maintain water-harvesting systems require attention (Table 5).

The water, energy and food balances in the Oum Zessar Watershed have been relatively well-studied. The total area of the aquifer is 1305 km^2 (Hadded et al., 2013). Average annual rainfall (209 mm) is split into evapotranspiration (72%), transmission loss (3%), percolation (19%) and outflow (6%) (Ouessar et al., 2009). Across most of the region, the recharge rate

Table 5. Water harvesting in the Zeuss Koutine Watershed and the Jordan River basin.

	Balance of ecosystem stocks and flows
Water	↑
Food and income	?
Natural cycling and storage of geochemicals	↑
Carbon sequestration and emissions reduction	↑

is 2.42% of rainfall (Hadded, 2008). But in the *wadis*, the rate of infiltration to groundwater recharge has been estimated at 70% of the total volume (10.230 MCM/y), due to the installation of the artificial recharge structures (Hadded et al., 2013).

The average energy consumption for urban water abstraction is estimated at 0.834 kWh/m^3 (Hadded et al., 2013). Since demand exceeds supply, part of the shortfall is made up by two desalination plants that have already been constructed to treat seawater, and further water treatment facilities are considered likely to be needed in future. These will use additional energy, and may emit waste-products.

Tunisia's second national communication to the UNFCCC (Tunisia, 2014) builds on the national adaptation strategy for agriculture (MARH, & GTZ, 2007) and those of other sectors. Scenarios for agricultural emissions, water supplies and productivity have all been developed, and action research to identify the effects of water harvesting and recharge practices is well integrated and supported through the National Agricultural Research and Extension System (WLI, 2013; WOCAT, 2013). Although the Tunisian policies, and particularly the proactive water-conservation and groundwater-recharge interventions implemented in Southern Tunisia, can be seen as relatively exemplary for the region, the national communication itself still explicitly underlines the need for further mainstreaming and integration amongst the sectorial agendas.

The Jordan River basin includes parts of Lebanon, Syria, Jordan, Palestine and Israel (Figure A3 in the supplemental online data). This includes the catchments of the West Bank that feed the mountain aquifer which is shared by Palestine and Israel (Chenoweth, 2011; Comair, Gupta, Ingenloff, Shin, & Mckinney, 2013; Hoff, Bonzi, Joyce, & Tielbörger, 2011; Mansour, Peach, Hughes, & Robins, 2012). In the West Bank, many people depend on livestock production for their living, but natural vegetation provides only 15% of livestock food needs, and only during winter and spring (Sholi, 2013). The remaining feed requirements for livestock and food for humans is met through crop production, which is mainly rainfed due to restrictions on groundwater extraction for irrigation. Transporting water and feedstuffs or moving animals from overgrazed areas to less degraded lands requires fuel and vehicles, in addition to other transaction costs. Where sufficient feedstuffs cannot be grown locally, they have to be imported at high cost from Israel to Palestine.

Constructing and maintaining water-harvesting structures requires the use of vehicles and fuel as well as labour (Sholi, 2013). Harvesting rainwater can make the difference between failure and success for rainfed barley production, avoid the need for irrigation of crops, and reserve scarce water supplies for other local uses by households or downstream flows. However, farmers in the West Bank are not necessarily willing to invest their time, land and labour in this practice unless they can see a clear economic case for doing so. They do not anticipate benefits from increased groundwater recharge, because they are not permitted to construct wells (Sholi, 2013).

Palestine has a Climate Change Adaptation Strategy and Programme of Action which have identified sectorial vulnerabilities, including those in the agricultural and energy sectors, as well as water security challenges (UNDP, 2011; UNDP/PAPP, 2009). The strategy and programme have underlined the untapped potential for energy measures to simultaneously deliver climate mitigation and adaptation benefits, particularly in situations of increased water scarcity as a result of climate change.

There is no joint climate adaptation strategy for the Jordan River basin countries, nor any framework for water and carbon emission accounting. However, Jordan's second national communication to the UNFCCC draws on available water-balance studies for portions of the basin in Jordan, including one for the Amman Zarqa Basin (GoJ, 2009).

At this level, recognition of the scope for investigation of the water–energy–food and climate trade-offs associated with groundwater recharge in water-harvesting systems is just beginning (WLI, 2014). However, the potential scope for policies enabling payment or other incentives for these watershed services is already attracting attention.

Discussion

Although in all six systems explored in this article rural communities have been practising agricultural water management strategies with potential to maximize benefits to water, food and energy resource management under climate change, the available evidence concerning the merits of these practices has rarely been sufficient to convince policy makers that they provide solutions worth investing in. Instead, decision makers are opting to pay rising costs for wastewater treatment and desalination. The overviews of water-balance studies presented in this article demonstrate the uneven current state of knowledge in the selected basins and sub-basin catchments, despite the availability of promising tools as demonstrated by Hadded et al. (2013) and Hoff et al. (2011). How then can implementation of the nexus approach be connected to the proposed four-dimensional assessment framework?

The qualitative approach taken in this article, integrating consideration of the water balance with inputs and outputs of energy and geochemicals through the water basin systems, advocates and invites further refinement of the approximated ratios of positive and negative factors presented so far. This would be compatible with emerging approaches to ecosystem accounting (UN, 2012, 2014; WAVES, 2014), as well as ongoing water resource accounting efforts (Karimi, Bastiaanssen, & Molden, 2013; Molden & Sakthivadivel, 1999). Resource accounting would enable more precise and effective weighing of the comparative values of benefits and trade-offs in the nexus by quantifying the stocks and flows of water, energy and geochemicals through the systems.

Using three or more different metrics for quantifying the effects of strategies and practices complicates quantification and weighing of trade-offs between sectorial objectives. The simple binary weighting and ratios used in this article give equal weight to stocks and flows of widely different magnitudes. This severely limits the effective weighing of trade-offs. Economic cost–benefit assessment is more convenient for informing decision makers because it translates the varying quantities of each of the inputs and outputs or benefits into monetary units that are directly comparable (TEEB, 2011; Pascual & Muradian, 2010). This presents a foregone conclusion regarding the optimal strategy, avoiding the burden of subjective decision making. However, economic assessments are only as good as the underlying data-sets quantifying the stocks and flows of each commodity.

The disaggregated four-dimensional approach is more intelligible for consideration and discussion by stakeholders than an aggregated value may be, and avoids other widely observed drawbacks inherent in economic assessment approaches. These include failure to take into account environmental externalities including pollution and resource degradation, failure to assign sufficient value to ecosystem function and services, use of discount rates that favour short-term gains over long-term sustainability (see further discussion in Chambwera et al., 2014; Pascual & Muradian, 2010) and general unintelligibility of the economic 'black box' amongst concerned stakeholders.

None of the countries where the case studies are located has yet adopted ecosystem accounting methods wholesale at the national scale, let alone sub-national or basin scales. Nevertheless, conventional agricultural and trade statistics (available through national statistical agencies, FAOSTAT and UNCOMSTAT) already capture some elements of the required accounting at a range of scales (i.e. crop production and inputs, including stocks and flows of nutrients and agrochemicals). Through national strategies in response to climate change (the fourth dimension of the water–energy–food–climate nexus), some of the missing elements required to enable ecosystem accounting tools to be used for quantitative basin-level assessment of trade-offs in the nexus are gaining increased attention.

The potential for agricultural activity to increase stocks of groundwater recharge and stocks of carbon through sequestration in soils and vegetation is not yet accounted for in any of the national strategies or accounting frameworks. However, necessary methods for carbon accounting are increasingly available and promoted through climate adaptation and mitigation efforts (Braimoh, 2012). In the meantime, national emissions accounts have been established in many countries of the region, providing the basis for a systematic approach to tracking the flows of carbon associated with agriculture and other sectors (UNFCCC, 2006).

Conclusions

In this article, the scope for four-dimensional hydro-eco-agrological assessments and improving nexus decision making has been illustrated through case studies in six selected basins. Underlying the case studies are varying levels of data availability concerning the water, energy and food stocks, flows and processes (major sources are indicated in the references). Some flows, such as inputs of climatic data and extents of vegetated areas, are increasingly well quantified, whereas others, such as groundwater extraction volumes and energy uses in the product cycle, remain more challenging to effectively quantify due to data-collection challenges. Adaptive management strategies should use the best available data and methodologies, including decision-support tools, to improve this situation through monitoring and assessment, coupled with the continued implementation of 'best-bet' green water management strategies. Providing resources, imperatives and review fora for this continuous learning-while-doing should be an essential element of nexus implementation policies.

The desirability and the challenge of achieving systematic quantitative assessments enabling more precise and effective weighing of nexus trade-offs associated with water management decision making has been explored through the qualitative assessment presented in this article. Discussion of the feasibility of progression from the qualitative assessment of nexus trade-offs to the systematic quantitative ecosystem accounting approach that is advocated focuses on the contribution of ongoing strategic work addressing climate change adaptation and mitigation as the fourth dimension of the water–energy–food–climate nexus. This ensures the progressive development of national frameworks for data generation and collation that will be key to the intended tracking of the other three dimensions (water, energy and food).

Although the focus of this article is primarily on assessment methods for policy formulation and improvement, rather than on advocating particular hardware fixes, the various patterns of livestock and human water uses illustrated have very different energy implications, and suggestions emerging e.g. in Palestine to give further attention to smallholders' access to sustainable energy sources and technologies heighten attention to the potential of this entry point for strategic intervention. Other recurring themes in the case studies offering potential entry points for nexus policies concern tightening of the internal efficiencies and feedbacks in recycling of water and nutrients in the production systems, including domestic, urban and industrial parts of the systems, and where produce is exported; use of systems to reward organic, sustainable or good agricultural production methods; and improved environmental monitoring.

Overall, the nexus approach is challenging but has significant potential to reveal, advocate and maximize the contribution of rural communities' green water management practices to the achievement of sectorial goals to cultivate a model of water, energy, food and climate security that will be accessible to the bottom billion in the drylands and help them maintain their quality of life. The case for increased strategic support for these practices appears stronger when weighed from the water–energy–food–climate nexus perspective, rather than purely from the point of view of water balance and food production.

Acknowledgements

The authors are grateful to Martin Keulertz, Eckart Woertz and the organizers of the international conference, The Water-Food-Energy-Climate Nexus in Global Drylands, for supporting the preparation of this article. We also wish to thank Dr Ahmed Al Falahi and several anonymous reviewers, who provided helpful and constructive reviews. The work by researchers at National Agricultural and Extension Systems referred to in the article was partially supported by USAID through ICARDA's Water and Livelihoods Initiative. Backgrounds of figures in the online supplementary material were derived from a base map available from Esri (http://goto.arcgisonline.com/maps/World_Topo_Map).

Disclosure statement

No potential conflict of interest was reported by the authors.

Supplemental data

Supplemental data for this article can be accessed at http://dx.doi.org/10.1080/07900627.2015.1026436

References

Al-Bakri, J., Suleiman, A., Abdulla, F., & Ayad, J. (2011). Potential impact of climate change on rainfed agriculture of a semi-arid basin in Jordan. *Physics and Chemistry of the Earth, Parts A/B/C, 36*, 125–134. doi:10.1016/j.pce.2010.06.001

Al-Charideh, A. (2013). Recharge and mineralization of groundwater of the upper cretaceous aquifer in Orontes basin, Syria. *Hydrological Sciences Journal, 58*, 452–467. doi:10.1080/02626667.2012.752578

Al-Falahi, A. (2013). *Water management strategies and impacts on livelihoods in Iraq* In A. ICARDA Water and Livelihoods Initiative (WLI) 5th Regional Coordination Meeting, Jordan, 2013 (Ed.) (p. 29). Amman: ICARDA.

Al-Falahi, A. (2014). *The Water and Livelihoods Initiative (WLI) Iraq country research review report 2010–2013*. Unpublished.

Ash, N., Blanco, H., Brown, C., Garcia, K., Henrichs, T., Lucas, N., & ... Zurek, M. (Eds.). (2010). *Ecosystems and human well-being a manual for assessment practitioners*. Washington, DC: Island Press.

Atroosh, K. B. (2013). *Strategic approaches to integrated management of land, water and livelihoods in the Abyan Delta, Yemen* In -.S. Regional Knowledge Exchange on decision Support Tools and Models, Djerba, Tunisia (Ed.) (p. 10). Amman: ICARDA.

Atroosh, K. B., & Moustafa, A. (2012). An estimation of the probability distribution of Wadi Bana flow in the Abyan Delta of Yemen. *Canadian Journal of Agricultural Science, 4*, 80–89.

Attaher, S. M., & Medany, M. A. (2008). *Analysis of crop water use efficiencies in Egypt under climate change*. Proceedings of the first international conference on Environmental Studies and

Research " Natural Resources and Sustainable Development", 7–9 April, Sadat Academy of Environmental Science, Minofya, Egypt.

Attia, F., Fahmi, H., Gambarelli, J., Hoevenaars, R., Slootweg, R., & AbdelDayem, S. (2005). *WDWCIARP drainframe analysis, main report.* Unpublished.

Braimoh, A. (2012). *Assessing the carbon benefits of improved land management technologies.* Retrieved from Report#: 68189: http://www-wds.worldbank.org/external/default/WDSContentServer/WDSP/IB/2012/04/20/000333038_20120420022123/Rendered/PDF/681890BRI00PUB0nagementTechnologies.pdf

Carpenter, S. R., Pingali, P. L., Bennett, E. M., & Zurek, M. B. (Eds.). (2005). *Ecosystems and human well-being: Scenarios* (Vol. Volume 2). New York, NY: Island Press.

Chambwera, M., Heal, G., Dubeux, C., Hallegatte, S., Leclerc, L., Markandya, A., & ... Zilberman, D. (2014). Chapter 17. Economics of adaptation. In IPCC (Ed.), *IPCC WGII AR5.* Cambridge, UK and New York, NY: Cambridge University Press.

Chenoweth, J. (2011). Will the water resources of Israel, Palestine and Jordan remain sufficient to permit economic and social development for the foreseeable future? *Water Policy, 13,* 397–410. doi:10.2166/wp.2010.131

Comair, G. F., Gupta, P., Ingenloff, C., Shin, G., & Mckinney, D. C. (2013). Water resources management in the Jordan River Basin. *Water and Environment Journal, 27,* 495–504. doi:10.1111/j.1747-6593.2012.00368.x

Comair, G. F., McKinney, D. C., Scoullos, M. J., Flinker, R. H., & Espinoza, G. E. (2013). Transboundary cooperation in international basins: Clarification and experiences from the Orontes river basin agreement: Part 1. *Environmental Science & Policy, 31,* 133–140. doi:10.1016/j.envsci.2013.01.006

EEAA. (2010a). *Egypt National Environmental, Economic and Development Study (NEEDS) for climate change under the united nations framework convention on climate change.* Cairo: EEAA. Retrieved from http://www.eeaa.gov.eg/English/reports/CCRMP/7.CCWaterStrategy/CCFinalSubmitted8-March2013AdptStrtgy.pdf

EEAA. (2010b). *Egypt second national communication under the united nations framework convention on climate change.* Cairo: EEAA. Retrieved from http://unfccc.int/resource/docs/natc/egync2.pdf

El-Qousy, D. A., Mohamed, M. A., Aboamera, M. A., & Kheira, A. A. A. (2006). On-farm energy requirements for localized irrigation systems of citrus in old lands. *Misr J. Ag. Eng., 23,* 70–83.

Evans, J. P. (2009). Twenty-first century climate change in the Middle East. *Climatic Change, 92,* 417–432. doi:10.1007/s10584-008-9438-5

Evans, R., Soppe, R., Barrett-Lennard, E., & Saliem, K. A. (Eds.). (2012). *Salinity in Iraq – Current state, causes, and impacts - an overview of the scope and scale of soil and water salinity problems in central and southern Iraq.* Amman, Jordan: ICARDA.

Falkenmark, M., & Molden, D. (2008). Wake up to realities of river basin closure. *International Journal of Water Resources Development, 24,* 201–215. doi:10.1080/07900620701723570

FAO. (2012). *Crop yield response to water.* Irrigation and drainage paper 66. Rome: FAO. Retrieved from: http://www.fao.org/docrep/016/i2800e/i2800e00.htm

Fraenkel, P. L. (1986). *Water lifting.* Rome. Retrieved from FAO Irrigation and Drainage Paper 43: Rome, Italy: United Nations Food and Agriculture Organization. Retrieved from http://www.fao.org/docrep/010/ah810e/ah810e00.htm

Gallopín, G. C. (2006). Linkages between vulnerability, resilience, and adaptive capacity. *Global nvironmental Change, 16,* 293–303. doi:10.1016/j.gloenvcha.2006.02.004

García, F., Díaz, J. A. R., Poyato, E. C., Montesinos, P., & Berbel, J. (2014). Effects of modernization and medium term perspectives on water and energy use in irrigation districts. *Agricultural Systems, 131,* 56–63.

GoJ. (2009). *Jordan's second national communication to the United Nations Framework Convention on Climate Change (UNFCCC).* Retrieved from http://unfccc.int/resource/docs/natc/jornc2.pdf

Hadded, R. (2008). *Update of the hydro geological model of the Zeuss Koutine aquifer and evaluation of the impact of water and soil conservation works on the recharge.* Dissertation, National Institute of Agronomy of Tunisia (INAT).

Hadded, R., Nouiri, I., Alshihabi, O., Maßmann, J., Huber, M., Laghouane, A., & ... Tarhouni, J. (2013). A decision support system to manage the groundwater of the Zeuss Koutine aquifer using the WEAP-MODFLOW framework. *Water Resources Management, 27,* 1981–2000. doi:10.1007/s11269-013-0266-7

Herren, H. (2011). *Agriculture - investing in natural capital*. Nairobi: United Nations Environment Program. Retrieved from http://www.unep.org/greeneconomy/Portals/88/documents/ger/GER_2_Agriculture.pdf

Hoff, H. (2011). *Understanding the nexus. Background paper for the Bonn 2011 conference: The water, energy and food security nexus*. Stockholm. Retrieved from http://www.water-energy-food.org/en/whats_the_nexus/background.html

Hoff, H., Bonzi, C., Joyce, B., & Tielbörger, K. (2011). A water resources planning tool for the Jordan River basin. *Water, 3*, 718–736. doi:10.3390/w3030718

Hussien, H. H., Aoda, M. I., & Alfalahi, A. A. (2014). The impacts of irrigation water salinity and deficit irrigation on sunflower helianthus annuus L. (in Arabic). *Iraqi Journal of Agricultural Research, 19*, 50–63.

IAASTD. (2009). *Agriculture at a crossroads - international assessment of agricultural knowledge, science and technology for development global report*. Washington, DC: Island Press.

ICARDA. (2013). *Brainstorming meeting on 'water and livelihoods in Yemen' organized for USAID by ICARDA 25th February, 2013 Movenpick Hotel, Sanaa*. Amman: ICARDA. Retrieved from http://www.icarda.org/wli/pdfs/summary_reports_yemen_meeting.pdf

IPCC. (2014). *Climate change 2014: Impacts, adaptation, and vulnerability*. Yokohama: IPCC. Retrieved from Assessment Report 5: http://www.ipcc.ch/report/ar5/wg2/

Jaafar, H. H. (2014). Feasibility of groundwater recharge dam projects in arid environments. *Journal of Hydrology, 512*, 16–26. doi:10.1016/j.jhydrol.2014.02.054

Jaafar, H. H. (2015). Impact of the Syrian conflict on irrigated agriculture in the Orontes. Basin. *International Journal of Water Resources Development* doi:10.1080/07900627.2015.1023892

Karimi, P., Bastiaanssen, W. G. M., & Molden, D. (2013). Water accounting plus (WA +) – A water accounting procedure for complex river basins based on satellite measurements. *Hydrology and Earth System Sciences, 17*, 2459–2472. doi:10.5194/hess-17-2459-2013

Keys, P., Barron, J., & Lannerstad, M. (2012). *Releasing the pressure: Water resource efficiencies and gains for ecosystem services*. Nairobi. Retrieved from http://www.sei-international.org/mediamanager/documents/Publications/Air-land-water-resources/sei-unep-releasing-the-pressure.pdf

Mansour, M. M., Peach, D. W., Hughes, A. G., & Robins, N. S. (2012). Tension over equitable allocation of water: Estimating renewable groundwater resources beneath the West Bank and Israel. *Geological Society, London, Special Publications, 362*, 355–361. doi:10.1144/SP362.20

MARH, & GTZ. (2007). *Stratégie nationale d"adaptation de l"agriculture tunisienne et des écosystèmes aux changements climatiques*. Retrieved from http://www.chm-biodiv.nat.tn/dmdocuments/economie_biodiv/snaatecc_cahier_4_27012007.pdf

Massaad, R., & Jomaa, I. (2013). *Water management strategies and impacts on livelihoods in Lebanon* In A. ICARDA Water and Livelihoods Initiative (WLI) 5th Regional Coordination Meeting, Jordan, 2013 (Ed.) (p. 32). Amman: ICARDA.

Mehari, A., Van Steenbergen, F., & Schultz, B. (2011). Modernization of spate irrigated agriculture: A new approach. *Irrigation and Drainage, 60*, 163–173. doi:10.1002/ird.565

MoE. (2011). *Lebanon's second national communication to the united nations framework convention on climate change*. Beirut: MoE.

MoE. (2012). *Lebanon technology needs assessment report for climate change*. Beirut, Lebanon. Retrieved from http://www.lb.undp.org/content/dam/lebanon/docs/EnergyandEnvironment/Publications/TNA_Book.pdf

Molden, D., & Sakthivadivel, R. (1999). Water accounting to assess use and productivity of water. *International Journal of Water Resources Development, 15*, 55–71. doi:10.1080/07900629948934

MWRI. (2010). *Draft strategy for water resource management until 2050 (in Arabic)*. Cairo: MWRI.

Neama, A. S., Al-Falahi, A. A., & Hamoudi, H. M. (2014). Response of eggplant (Solanum melongena L.) to subsurface drip irrigation method in comparison with surface drip and furrow irrigation methods under conventional agriculture (in Arabic). *Iraqi Journal of Agricultural Research, 19*, 221–215.

Nour El-Din, M. (2013). *Climate change risk management in Egypt proposed climate change adaptation strategy for the ministry of water resources & irrigation in Egypt prepared for UNESCO-Cairo office*. Cairo: EEAA. Retrieved from http://www.eeaa.gov.eg/English/reports/CCRMP/7.CCWaterStrategy/CCFinalSubmitted8-March2013AdptStrtgy.pdf

Ouessar, M., Bruggeman, A., Abdelli, F., Mohtar, R. H., Gabriels, D., & Cornelis, W. M. (2009). Modelling water-harvesting systems in the arid south of Tunisia using SWAT. *Hydrology and Earth System Sciences, 13,* 2003–2021. doi:10.5194/hess-13-2003-2009

Ouessar, M., Sghaier, M., Mahdhi, N., Abdelli, F., De Graaff, J., Chaieb, H., & ... Gabriels, D. (2004). An integrated approach for impact assessment of water harvesting techniques in dry areas: The case of Oued Oum Zessar watershed (Tunisia). *Environmental Monitoring and Assessment, 99,* 127–140. doi:10.1007/s10661-004-4013-7

Pascual, Unai, & Muradian, R. (2010). Chapter 5: The economics of valuing ecosystem services and biodiversity. In *The economics of ecosystems and biodiversity: The ecological and economic foundations.* The Economics of Ecosystems and Biodiversity (TEEB). London and Washington: Earthscan.

Pretty, J. (2003). Agroecology in developing countries: The promise of a sustainable harvest. *Environment: Science and Policy for Sustainable Development, 45,* 8–20. doi:10.1080/00139150309604567

Pretty, J. N., Noble, A. D., Bossio, D., Dixon, J., Hine, R. E., Penning de vries, F. W. T., & Morison, J. I. L. (2006). Resource-conserving agriculture increases yields in developing countries. *Environmental Science & Technology, 40,* 1114–1119. doi:10.1021/es051670d

Rockström, J. (1999). On-farm green water estimates as a tool for increased food production in water scarce regions. *Physics and Chemistry of the Earth, Part B: Hydrology, Oceans and Atmosphere, 24,* 375–383.

Rockström, J., & Gordon, L. (2001). Assessment of green water flows to sustain major biomes of the world: Implications for future ecohydrological landscape management. *Physics and Chemistry of the Earth, Part B: Hydrology, Oceans and Atmosphere, 26,* 843–851.

Rockström, J., Karlberg, L., Wani, S. P., Barron, J., Hatibu, N., Oweis, T., & ... Qiang, Z. (2010). Managing water in rainfed agriculture – The need for a paradigm shift. *Agricultural Water Management, 97,* 543–550.

RoY. (2013). *Republic of Yemen second national communication under the United Nations Framework Convention on Climate Change.* UNFCCC. Retrieved from http://unfccc.int/resource/docs/natc/yemnc2.pdf

Safriel, U., Adeel, Z., Niemeijer, D., Puigdefabregas, J., White, R., Lal, R., & ... McNab, D. (2005). Dryland systems. Millennium ecosystem assessment. *Current state and trends* (Vol. 1, pp. 623–662). Washington, DC: Island Press.

Saleh, E. M. A., Ibrahim, A. A., Dhehibi, B., & Hassan, A. A. (2013). Economic assessment of some technologies used in irrigated agriculture and their impact on farmers' livelihoods: case of the Egyptian salt-affected soils Farms. *Egyptian Journal of Agricultural Economics, 23*(3).

Sghaier, M. (Ed.). (2009). *Proceedings of the Colloque International Sociétés en transition et développement local en zones difficiles "DELZOD".* Institut des Régions Arides, (Laboratoire d'Economie et Sociétés Rurales, LESOR), Médenine – Tunisie 22–24 Avril 2009.

Sholi, N. (2013). *Water management strategies and impacts on livelihoods in Palestine* In R.K.E.o. D.-s. T.a.M. T.-S. 2013 (Ed.) (p. 30). Amman: ICARDA.

TEEB. (2011). *TEEB manual for cities: Ecosystem services in urban management.* Retrieved from http://www.teebweb.org/publication/teeb-manual-for-cities-ecosystem-services-in-urban-management/

TR. (2011). *Turkey's national climate change adaptation strategy and action plan.* Ankara: TR. Retrieved from http://iklim.cob.gov.tr/iklim/Files/Belgeler/NationalAdaptationStrategy.pdf

Tunisia. (2014). *Seconde Communication Nationale de la Tunisie à la Convention Cadre des Nations Unies sur les Changements Climatiques.* Retrieved from http://unfccc.int/resource/docs/natc/tunnc2.pdf

UN. (2012). *International recommendations for water statistics.* New York, NY: UN. Retrieved from Statistical papers Series M No. 91: https://unstats.un.org/unsd/envaccounting/irws/irswebversion.pdf

UN. (2014). *System of environmental-economic accounting 2012 central framework.* E. Union. Retrieved from http://unstats.un.org/unsd/envaccounting/seeaRev/SEEA_CF_Final_en.pdf

UNDP. (2013). *Water governance in the Arab region - managing scarcity and securing the future.* Republic of Lebanon: Retrieved from http://www.undp.org/content/dam/rbas/doc/Energy and Environment/Arab_Water_Gov_Report/Arab_Water_Gov_Report_Full_Final_Nov_27.pdf UNDP.

UNDP, P. (2011). *Climate change adaptation strategy and programme of action for the Palestinian Authority*. Retrieved from http://www.undp.ps/en/newsroom/publications/pdf/other/climatechange.pdf

UNDP/PAPP. (2009). *Climate change adaptation strategy for the occupied Palestinian territory*. Jerusalem: UNDP/PAPP.

UNFCCC. (2006). *Updated UNFCCC reporting guidelines on annual inventories following incorporation of the provisions of decision 14/CP.11 Note by the Secretariat*. Subsidiary Body for Scientific and Technological Advice Twenty-fifth session Nairobi, 6–14 November 2006, Item 7 (b) of the provisional agenda, Methodological issues under the Convention, Issues relating to greenhouse gas inventories. Retrieved from http://unfccc.int/resource/docs/2006/sbsta/eng/09.pdf

Voss, K. A., Famiglietti, J. S., Lo, M., De Linage, C., Rodell, M., & Swenson, S. C. (2013). Groundwater depletion in the Middle East from GRACE with implications for transboundary water management in the Tigris-Euphrates-Western Iran region. *Water Resources Research, 49*, 904–914. doi:10.1002/wrcr.20078

Watson, R. T., & Zakri, A. H. (2005). *Living beyond our means: Natural assets and human well-being: Statement from the board*. Washington, DC: World Resources Institute.

WAVES. (2014). *Users and uses of environmental accounts: A review of select developed countries*. Washington, DC: WAVES. Retrieved from https://http://www.wavespartnership.org/sites/waves/files/documents/PTEC1-UsersandUsesofEnvironmentalAccounts.pdf

WEF. (2014). *Call to action*. Bonn: WEF. Retrieved May 20, 2014, from http://wef-conference.gwsp.org

WLI. (2013). *Improved use of scarce water resources to adapt Tunisian livelihoods to climate change*. Tunis: ICARDA.

WLI. (2014). *Water and livelihoods initiative, annual report, 2013*. Amman, Jordan: WLI. Retrieved from http://www.icarda.org/wli/pdfs/Anual-Report-FinalFebruary3-2013.pdf

WOCAT. (2013). *Water harvesting – Guidelines to good practice*. Berne: CDE.

WWDR. (2014). *Water and energy*. Paris: UNESCO.

The nexus as a political commodity: agricultural development, water policy and elite rivalry in Egypt

Harry Verhoeven[a,b]

[a]Department of Politics & International Relations, University of Oxford, UK; [b]School of Foreign Service (Qatar), Georgetown University, Washington, DC, USA

Thinking of the interconnections between water, food, energy and climate is nothing new in the Nile Basin; it has long been anchored in political struggles. For 200 years, Egypt's political economy has been defined by water use patterns and food security strategies that debunk the technocratic myth that rapid growth, interaction with global markets and technological modernization eliminate poor governance practices and allocative inefficiencies. In contrast, the prism of the nexus as a political commodity illuminates one of modern Egypt's most consequential dialectics: the interaction between the very particular nexus at the heart of the country's political economy, forged through factional strife and sustained by outside discourses and interests, and the economic and ecological ravages of this elite politics. Egyptian history serves as a warning. Today's conversation needs to be deconstructed in terms of how different forms of interconnectivity between water, energy and food are produced and experienced by different social groups. It reminds us to take interconnections not as given, but rather as contested and contestable outcomes from which opportunities for adaptation and transformation do not naturally emerge, but need to be struggled for.

Long-term environmental trends like deforestation and acute catastrophes like Hurricane Katrina in 2005 and the 2011 Fukushima nuclear disaster feature prominently on the global agenda of the early twenty-first century. Historically, worries about the environment as a threat to human civilization have been cyclical, as has been the framing of man's ability to respond effectively to such a threat (Fleming, 2010). The salience of environmental narratives has depended on geopolitical imperatives and economic cycles, yet strikingly in the last two decades (despite the Great Recession), concerns about global warming, commodity prices and energy security have not faded away. For an increasing number of observers these developments are an integral part of a wider crisis of the global economic system (De Schutter, 2011; Patel, 2007).

The tendency not to see environmental issues in isolation from each other and not to separate them from social and economic problems has partly been the outcome of environmentalists trying to mainstream their cause and convince more anthropocentric constituencies to take ecological questions seriously; and partly it is the result of theoretical scholarship and empirical evidence highlighting the interconnections between different spheres of human activity and their intersection with the environment (Conca, 2006). The 1992 United Nations Conference on Environment and Development and the 2002 World Summit in Johannesburg provided momentum for highlighting the linkages between water management, agriculture and industrialization; health and pollution; sustainability and power generation; etc. In the run-up to the Rio + 20 Conference on Sustainable Development, most preparatory working groups and reports emphasized water as the central node connecting environmental, social and economic dynamics: "Water security is the gossamer that links together the web of food, energy, climate, economic growth, and human security challenges that the world economy faces over the next two decades" (Waughray, 2011, p. 1). North of 80% of globally available water is used for agricultural production; increasing productivity will require more efficient and less (economically and environmentally) costly irrigation, including through the electrification of pumping. Concurrently, hydropower provides about 20% of global electricity, with China and Brazil being the world's biggest producers of hydroelectric power in 2013; in the latter, almost three-quarters of electricity generation is through water.

This ostensibly comprehensive approach has been termed the 'nexus' (from the Latin *nectare*, to bind together) to highlight the hard-wired linkages between the supply of and demand for water, energy and food (and, more recently, climate; Bazilian et al., 2011). This approach does not merely refer to growing environmental pressures and the interdependence of ecosystems but also explicitly casts itself as 'solution-oriented' (see e.g. Asian Development Bank, 2013). The hitherto dominant framing of what the nexus is – in its definitions of interrelated problems and interconnected solutions as well as in the proposed methodology of inquiry and implementation – represents a logical extension of the mainstream paradigm of sustainable development, which in the water sector is conceptualized as integrated water resources management (IWRM).

This approach has three main characteristics: first, a positivist understanding of environmental change and man's role in affecting that change; second, a focus on institutions – international treaties, markets, transboundary networks, etc. – which are assessed as to how helpful they are in impacting the nexus; and third, a deliberately apolitical identity. While the combination of these characteristics gives 'the nexus', like sustainable development (Lélé, 1991) or IWRM (Jensen, 2013), a high degree of plasticity to allow it to mean different things to different people, such an omnivorous conceptualization also veils that there is not one single nexus but multiple, socially constructed and politically consequential nexuses. Otherwise put: the idea of a global web of connections does not tell us much about the origins and distributional significance of certain sets of links between water, food and energy (and the absence of others) at particular scales; nor does it tell us much about what role power plays in framing the nature, intensity and frequency of the connections between the constituent factors.

Suggesting that the environment is political is hardly radical or novel, of course. Political ecology has anatomized how our ways of seeing the 'natural' cannot be isolated from our organization of human societies and how labels such as 'fragile', 'manageable' or '(un)sustainable' are never politically agnostic. Underscoring the interaction between environmental knowledge, power struggles and social values enables us to bring the rich analytical arsenal of class, scale, discourse and gender to bear on the depoliticizing tendencies of both technocracy and Malthusianism to evince how power manifests itself, directly and indirectly, in our changing environments and the storylines that accompany them. Policy shifts, like paradigmatic transformations in science, do not emerge from the rational accumulation of carefully weighed and neutrally judged evidence but must be understood in terms of their historical context and the coalitions that coalesce and fragment around particular discursive and material configurations (e.g. IWRM – see Allan, 2003). Through adopting particular lenses and scripts to read, control and govern environmental processes, elites reshape communities, markets and states (Watts, 2004).

But these attempts have variable success rates, as social constructions of the environment reflect the political, but are entirely co-constitutive of it too; both power holders and their critics often overestimate the state's ability and underestimate the relative autonomy of local dynamics (Migdal, 1988).

The nexus as control over people

According to UNESCO director-general Irina Bolkova's introduction in the latest UN World Water Development report, "The fundamental right to freshwater is not exercised by some 3.5 billion women and men" (United Nations World Water Assessment Programme, 2014, p. xv). By 2050, one-quarter of the global population will live in a country affected by chronic freshwater deficits. The water–energy–food–climate nexus is of particular significance in global drylands (Koohafkan & Stewart, 2008), a category in which Africa's deserts, savannahs and Sahel have long occupied a special place for policy makers and scholars. No ecological zone has been more subjected to externally inspired environmental crisis narratives – essential fellow travellers of the colonial project and post-independence visions for authoritarian state-building by African elites (Verhoeven, 2014) – than Africa's drylands (Gibson, 1999). This history of the entwining of politics and environmental policies, often approached through the prism of antagonism between local communities and their habitats (Anderson & Grove, 1987), cannot be ignored. The importance of not merely including a historical dimension to debates around 'the nexus', but understanding that any nexus is inherently political, allows for a more complex, multi-layered grasp of the institutions that shape the interconnectivity between flows of water, energy and food. This means going beyond one-dimensional conceptions of how the environment and the political intersect – e.g. the unbearable lightness of the prophesied Water Wars (De Stefano, Edwards, de Silva, & Wolf, 2010) or the IWRM buzz of better governance which has little to say about accumulation, scalar effects or subversion (Bakker & Morinville, 2013).

Rather than assuming a unidirectional relationship in which global forces remake the non-Western world, I emphasize political agency to understand how local elites leverage transnational economic and environmental trends and storylines for their own strategies of state-building and/or domestic political consolidation. This enables more fruitful explorations of the distributional effects of deepening linkages and the complexities of operationalizing global agendas into local strategies, e.g. with regard to adapting to climate change or boosting agricultural productivity (Mortimore, 2010). From this perspective, nexus discourses arise from particular political-economic complexes, and the narrative of the nexus – like that of sustainable development, resource scarcity or water wars – can be deployed as a commodity in political struggles, to marginalize opponents, to exclude peripheral populations, to obscure allocation patterns.

Interpreting the nexus as a political commodity highlights the ability of local leaders to instrumentalize and, ultimately, claim ownership of yet another, externally produced crisis narrative that worries about African (and other non-Western) environments and advances intrusive policy prescriptions. For incumbent regimes, the interrelated challenges of water scarcity, food supply and climate change are, politically, merely one more set of material problems and one more internationally salient discourse as they navigate the "politics of permanent crisis" (Van De Walle, 2001). The exceptional character that many nexus scholar-activists attribute to it is not experienced as such by local elites; the nexus is yet another trope that ruling factions in the peripheries of the global system evaluate in terms of its potential for resource extraction internationally and dominance domestically (Verhoeven, 2011).

The overarching political dynamic to contend with is that of the central role of elite rivalries and how this shapes water, energy and food policies and practices – and the interrelations between them. Shifting coalitions in limited-access orders (North, Wallis, & Weingast, 2009) employ a wide spectrum of legal, extralegal and illegal tactics to structure patterns of rent-seeking to their advantage; shrewd narratives, patronage and violence are all parts of the toolkit. While technocratic approaches maintain the fiction that nexus-related policies are ends in themselves and judge them as failures or success on these terms, elites in limited-access orders subjugate these almost entirely – if not always successfully, because of the relative autonomy of economic and ecological dynamics – to the objectives of survival and relative gains in the balance of power. Rent-seeking is not merely an economic nuisance. In rural societies with a high dependence on irrigation provision or agricultural inputs, it is a highly effective political strategy of tying various constituencies to those in power through water, electricity and land and creating an asymmetric interdependence between patrons and clients (Wade, 1982). While these rents provide some degree of redistribution, they underlie the brutish reproduction of agrarian power blocs. Analyzing 'the nexus' as a political commodity captures the instrumentalization of global discourses at the local level to legitimize or challenge authority structures and also refers to the material practices that underpin the power exercised by ruling classes through exclusionary hydropolitical economies.

In the remainder of this article, I use the example of Egypt to challenge orthodox accounts which assume that adopting a nexus frame is politically benign and therefore desirable. It is not the absence of nexus logic that explains the persistence of hunger, water scarcity and authoritarianism in Cairo. Thinking of the interconnections between water, food, energy and climate is nothing new in the Nile Basin; it has long been anchored in political struggles. For the last 200 years, Egypt's political economy has been defined by water use patterns and food security strategies that debunk the technocratic myth that rapid growth, interaction with global markets and technological modernization eliminate poor governance practices and allocative inefficiencies. In contrast, the prism of the nexus as a political commodity illuminates one of modern Egypt's most consequential dialectics: the interaction between the very particular nexus at the heart of the country's political economy, forged through factional strife and sustained by outside discourses and interests, and the economic and ecological ravages of this elite politics. Egyptian history serves as a warning. Today's conversation needs to be deconstructed in terms of how different forms of interconnectivity between water, energy and food are produced and experienced by different social groups. It reminds us to take interconnections not as given, but rather as contested and contestable outcomes from which opportunities for adaptation and transformation do not naturally emerge, but need to be struggled for (Ribot, 2014).

Everything changes so everything stays the same: two centuries of producing and reproducing an Egyptian nexus

One of the most pervasive myths about the Nile Basin is that of the timelessness of the dominant development model: the reverie of modernization through large-scale irrigation, buttressed by a powerful, centralized state controlling the allocation of water and investing in big hydro-infrastructure to green the riverbanks (Ayubi, 1980). Cairo's rulers have long argued, domestically and internationally, that the structure of Egypt's political economy is the inevitable consequence of a unique geography – large-scale habitation is only possible along the Nile and in the Delta, and feeding the masses can solely be done through centrally led irrigated production – and thus the logical continuation of historical patterns of development (Mitchell, 2002). Moreover, history has been invoked not just to justify the contemporary order but also to pinpoint the dream all Egyptians are supposed to contribute to: that the splendours of millennia of Egyptian civilization were undergirded by an agrarian economy that was the Eastern Mediterranean's breadbasket and that reassuming the role of regional hegemon therefore requires readopting well-known recipes – regardless of whether wayward peasants agree.

Compelling as this storyline has sounded to international interlocutors, it is a modernist fiction. The dominant development model emerged only with the rise to power of Muhammad Ali Pasha and the project of modern state-building in the Nile Basin as a whole, kick-started from Cairo. Influenced by French *savants* who had brought the aspirations and illusions of the French Revolution to Egypt, Ali's vision was dominated by an imagined nexus between centralizing government, augmenting the water supply and boosting agricultural production (Fahmy, 2002). Ali believed that Egypt could throw off the Ottoman yoke and become a nineteenth-century Mediterranean power on a par with France or Britain only if it reformed its traditional, lethargic economy which was dependent on the caprices of the Nile and the climate. Industrialization and international trade would bring bullion and new technologies to Egyptian shores, but they necessitated an agrarian revolution: command-and-control agricultural production, with land concentrated in the hands of the ruler, and allocations of water (as well as decisions regarding cropping patterns) in those of his savants (Cuno, 1992).

Ali's state-building gamble was meant to quash domestic opposition and insulate him against foreign pressures. It presented itself as a rediscovery of the Pharaonic heritage: the image of the Pasha as the Farmer-General symbolized the entanglement of political power, legitimate authority and agricultural success through regimented planning. However, the economic strength of ancient Egyptian civilization had little to do with agricultural dirigisme and a politically crafted (and militarily imposed) nexus between a hydraulic mission of building large-scale infrastructure and boosting crop productivity (Eyre, 1994; Eyre, 1999). The Pharaohs did not dam the Nile, nor did administrators in Memphis regulate production on every *feddan* of land. Irrigation on a substantial scale was introduced only after 300 BC in Ptolemaic Fayum, and tenure and working conditions were, for most of antiquity, a far cry from the corvée labour Ali enforced (Butzer, 1976). Ancient Egypt's resilience stemmed from the decentralized character of food production and water management; that producers were allowed to experiment and flexibly adapt to fickle ecological conditions was the backbone of its wealth and regional influence.

When Muhammad Ali died in 1849, his command-and-control model, producing long-staple cotton for global markets, was encountering its economic and ecological limits. But Cairo's international standing had risen, and a military-commercial complex used the levers of the state to enrich itself and consolidate power (Vatikiotis, 1985). Ali's successors inherited his regional empire, nascent industries and – perhaps most importantly – the Pasha's ideas of what exercising power in Egypt meant and what the national destiny was. Dynastic rule thrived for another century. Successive Khedives, lacking the political strength of the dynasty's founder, relied on a land-owning oligarchy which they tied to them through mercantile favours and allocating thousands of *feddan* for crop production, in exchange for support for the monarchy's policy of Westernization and integration with global markets (Owen, 1993). During the American Civil War, cotton prices soared exponentially, and the area under cotton cultivation tripled in about a decade, leading to a bonanza for the royal family and its acolytes – a mixture of economic elites and (former) military commanders who were rewarded for their services. There was nothing *natural* about the political economy of the Alawiyya dynasty and the roles it attributed to labour, land, capital, water and food, although the combined rhetoric of a reconnection with Egypt's glorious past and the inescapable need to Westernize incessantly sought to naturalize the character of the regime and the nexus on which its hegemony depended (Mitchell, 1988).

Muhammad Ali's bequest was both a political imaginary and a material reality of class formation that continues to be integral to the reproduction of power and to the linkages between water, energy and food that underpin this process today. On the one hand, Egypt became a global agricultural player and was hailed as a beacon of modernity – home to Cairo's stunning opera house, the cosmopolitan *entrepôt* of Alexandria and the Suez Canal (Haddad, 2005). Khedive Ismail famously commented after the latter's completion that "My country is no longer in Africa; we are now part of Europe. It is therefore natural for us to abandon our former ways and to adopt a new system adapted to our social conditions" (de Guerville, 1906, p. 94). Francophile elites in Cairo hosted statesmen and empresses in garden palaces by the Nile, and Egyptian cotton was invaluable for that great trailblazer of industrialization, the textile mill.

On the other hand, there was the other Egypt: that of the Saeed, the country's unruly South, which periodically erupted in revolt and failed to play its prescribed role in the Alawi hydro-agricultural transformation (Brown, 1990b); the Egypt of the vast majority of rural dwellers, many of whom tried to escape the gruelling labour regime; and the Egypt of the 'traditional villages', supposedly frozen in time and untouched by modernity, waiting to be civilized (from the twentieth century onwards, 'developed'; Magd, 2010). Egypt's rulers and its international partners felt no need to deny the presence of this other, second Egypt. Its existence helped legitimize enlightened authoritarian patterns of rule, primitive accumulation and the penetration of the state in the "brigandist" Saeed (Brown, 1990a).

It was this political dyad of the Two Egypts that the 1952 Free Officers coup yearned to overturn, promising a wave of modernization that would 'finally' realize Egypt's manifest destiny. On the eve of Gamal Abdel Nasser's self-declared revolution, the dominant yet internally feuding power bloc – a military aristocracy, absentee landlords, exporters of cotton and sugar, and the king at its apex – owned as much as three-quarters of all cultivated land, with around two-thirds of economic assets owned by about 5% of the population (Osman, 2010, pp. 37–44). The Egyptian economy delivered superbly for them, as it did for foreign interests, particularly since the Suez Canal and the mounting debts of the late nineteenth century led to Egypt's incorporation into London's imperial orbit (Abdel-Fadil, 1975). However, the expansion of public education equipped the lower middle class not just with rising expectations but also with the organizational abilities to launch an offensive against the ossified *ancien régime*. Nasser vowed to make Egypt great again; the inclusion of the entire nation in his development strategy and the termination of foreign overrule were intimately connected (Baker, 1978). The new order prioritized industrialization instead of the capital-intensive agricultural sector and nationalized the Suez Canal. Together with the dismantling of the fledgling bourgeois democracy, this seemed to warrant the death knell for the practices and ideas of the previous 150 years.

Unfortunately, the Nasser regime's actual track record was more muddled and marked by continuities with Muhammad Ali's hegemonic strategy than many have cared to admit. Take for example the Free Officers' most illustrious policies: the Aswan High Dam and land reform. The former, the largest infrastructure project in Africa, was intended as a symbol of self-sufficiency, technical prowess and national sovereignty, but it drew on a similar conception of the connections between Egypt's geography, water supply and agriculture as those developed by Muhammad Ali – and added the Leninist notion of the electrification of the countryside. Nasser's water–energy–food nexus emerged from the same wellspring of high modernism as analogous pushes in the Soviet Union and United States and was an integral part of his attempts at liquidating rival elites in rural Egypt by weakening the control of the landowning class over the peasants. Aswan was a Farmer-General-style project of pulling Egypt into modernity by overcoming the curse of the unpredictable flood. It aimed to be simultaneously paradigmatic – as the icon of what Nasserite development entailed – and materially transformative of the country's web of internal and external relations.

The question of land ownership, too, would end up as a relatively minor variation on older melodies. While a legal cap of 200 *feddan* per individual was decreed to obliterate the huge estates along the Nile and in the Delta, the nucleus of the Egyptian political economy remained untouched. Some powerful families did lose their lands, but others found ways of circumventing the reforms and struck pragmatic deals with a government beset with paranoia about internal enemies and outside threats. Fears of rural insurrection, or an alliance between the old feudal-dynastic factions and foreign interests, led not to a Stalinist liquidation of commercial farmers but to their re-empowerment (Waterbury, 1983). The Free Officers increasingly relied on a rural, capitalist middle class, which was the logical counterpart of the new bureaucratic bloc that benefited from import-substitution industrialization policies. "The prerevolutionary social structure came to influence deeply the disposition of land and other benefits of the reform.... Thus society transformed the state" (Migdal, 1988, p. 185).

Another key reason that a real social transformation was thwarted lay in the power struggle inside the regime. The fierce competition between Nasser and Field Marshall Abdel Hakim Amer incentivized each protagonist to expand his respective patronage networks to offset the other's (Cook, 2012). Officially best friends, Nasser and Amer undermined each other relentlessly on the battleground of the political economy (Kandil, 2012). With the Nile as a foreign policy dossier in the hands of the presidency and the General Intelligence Directorate, Amer's power base was the military, which claimed a prominent position in the economy as senior officers accumulated large tracts of land, ran the foremost political offices and agro-businesses, and headed a state-within-the-state which Nasser did not dare dismantle (Hashim, 2011). This logic of competing clientage systems was diametrically opposed to the meritocratic, pro-smallholder policies that land reform supposedly fast-tracked (Adams, 1986). However, Nasser and Amer, in their desire not to lose the battle for political supremacy, relied on tactics of working with local strongmen and using land and water as patronage instruments. The price was the subversion of the Free Officers' initial revolutionary objectives.

The failed 1960s reforms exemplify how economic and environmental policies are subjugated to the logic of a limited-access order where elites constantly jockey for position and how various incarnations of what is today termed 'the nexus' have their origins in political rivalries. In turn, the Egyptian example also clearly illustrates the limits of what political voluntarism can achieve and how policy failure (from a technocratic nexus perspective) is 'irrationally' carried over from one generation to the next. Consider, for instance, agricultural productivity's trajectory since the nineteenth century when, following Muhammad Ali's big push and the cotton boom of the 1860s, myriad environmental problems eroded the output gains of increased land use (Richards, 1982). British colonialism from 1882 onwards re-energized the bid for an agricultural revolution through mass investment in rehabilitation and the practice of perennial irrigation (O'Brien, 1968). The objective was to get Egypt to repay its international debts and to provide cheap inputs to British industry – for neither the first nor the last time in Egyptian history, a particular nexus was crafted to serve overseas interests. However, the initial achievements under London's proconsul, the Earl of Cromer, were undermined in subsequent years by rising soil salinity and woeful drainage. Although autonomy from the Nile flood, thanks to the first Aswan Dam, lifted production spectacularly, the costs of mono-cropping and the obsession with increasing water supply were waterlogging, rampant bilharzia and decreasing crop quality. The economy as a whole, so dependent on cotton, barely grew as a result of policies which made political sense (e.g. the stability-ensuring partnership between the monarchy, British colonialism and wealthy landowners) but added less than 0.5% annually to national income per capita between 1900 and 1950 (Hansen, 1991, pp. 3–4, 39).

Nasser's reforms stimulated the economy generally, and the agricultural sector especially, as cotton lost much of its exalted status, smallholder farming was boosted and improvements in human capital raised productivity. The best decade of the twentieth century was the 1950s, when the emphasis on export agriculture declined and rice and cotton were less favoured than other cereals and animal feed cultivation. But as the import-substitution industrialization strategy became ensnarled in bureaucratic obstinacy and corruption, the arms race with Israel devoured resources, and the Amer–Nasser rivalry corroded the agricultural reforms, Egypt's political economy in the 1960s re-bred the ailments of old. The Aswan High Dam magnified the environmental and health problems that had bedevilled agricultural practices since the British built the first dam in 1902. Productivity dropped again and would barely recover under Nasser's successor, Anwar Sadat (Faris & Khan, 1993).

The Free Officers' nexus strategy left a final disastrous legacy, again born from its high modernist mindset and governance-through-patronage approach: the mirage of land reclamation. The bulk of the agricultural budget in the 1960s flowed to the Western Desert, where hugely expensive operations like the Tahrir project were undertaken to cultivate an additional 20% of land, as the Cairo government, as part of Egypt's hydraulic mission, sought to green the desert in a new bid for self-sufficiency in food grains. Perhaps only 33% of the reclaimed *feddan* were ever brought into production, with abysmal cost–benefit ratios. Economic and environmental failure did, however, respond to a clear political rationale as both Nasser and Amer wanted sweeping patronage opportunities of allocating lands to their respective clients (Springborg, 1979, 1982). Unsustainable as the policy was, its political importance sustained it, and the discourse of Egypt's heroically fighting the desert to alleviate poverty and environmental scarcity proved popular with Arab and (later) Western investors who partnered with the generals.

Much of the literature characterizes the Nasser-Amer era as one of leftist statism and that of Sadat and Hosni Mubarak (1981–2011) as neoliberal. While the private sector was certainly given greater space by the latter two administrations, this difference should not be overstated in light of the mixed public-private strategies all Egyptian presidents have used, with clientage as their central operating mechanism (Weinbaum, 1985). Moreover, the ideas inherited from Muhammad Ali about Egypt's manifest destiny and the key role for a water–food–climate nexus in building political coalitions have been discernible in the policies of every republican Pasha. What did shift were the concrete expressions of that logic and the specific role of economic actors in the nexus – as a function of the changing sociology of the regime and of the incumbent's calculus.

Both Sadat and Mubarak came to power as consensus figures without much of a military or economic power base of their own. Despite being army men, both sought to loosen their privileged ties to the institution that had raised them and to diversify their support networks to gain autonomy and protect themselves against a coup. Both presidents allied themselves to the business sector. Sadat embarked on his *infitah* ('openness') strategy to stimulate private investment; Mubarak would bring dozens of entrepreneurs, industrialists and former officials from the international financial institutions into the cabinet – and the security services. Egypt became more of a police state and less of a military junta. State security acquired military capabilities, grabbed important chunks of the economy and handled the Nile dossier for Mubarak, while many business figures and agro-industrial patrons moved into politics and had strong ties to the interior ministry, which was the first line of riot control in a context of growing inequality (Kandil, 2012).

The shrewdly crafted (and regularly reworked) connections between water, energy and food remain critical to the Egyptian political economy, premature obituaries notwithstanding. Despite endless development assistance missions to repair agricultural productivity, despite the decreed switch from cotton to cereal production, and despite the billions spent on land reclamation by Nasser, Sadat and Mubarak, the import bill keeps mounting, rural poverty is deepening, and environmental problems refuse to disappear. Throughout the 1970s and 1980s, total factor productivity in agriculture was probably lower than it was in 1900, and agriculture's contribution to growth was minimal (Hansen, 1991). Today no country in the world imports more wheat and, indirectly, more water than Egypt (Allan, 2011).

Against the backdrop of a rising population – today there are 40 times as many Egyptians as in Muhammad Ali's days – riverain, capital-intensive agriculture lost its lustre under Sadat and Mubarak as cotton's importance to the global economy wilted. Egypt has been increasingly unable to compete, losing the international shine it enjoyed during parts of the nineteenth and mid-twentieth centuries. The root cause of failure was and is political: the contradictions assembled as a result of a political survival strategy domestically, a function of the structural dilemmas facing elites in limited-access orders. Sadat's *infitah*, the 1978 peace deal with Israel and the alliance with Washington – a three-pronged course maintained by Mubarak under the banner of stability – produced a number of interrelated results that characterized a political economy of stagnation (Richter, 2007). Sadat's and Mubarak's arm's-length relationship with the armed forces meant that the latter needed to be kept out of domestic politics and foreign policy (Harb, 2003). While the USD 1.3 billion in annual assistance gave the soldiers plenty of hardware to play with, the Camp David agreements ended their permanent mobilization and the ability of individual officers to emerge as heroes (and thus political threats to the incumbent; Kandil, 2011). The generals were pushed further into business and progressively marginalized in the cabinet to wed them, grudgingly, to the status quo. Unhappy as they were with their loss of pre-eminence, their involvement in patronage networks meant that many of them stood to lose from more radical change in the form of liberal democracy or Islamism. Thus, the language of accountability and free markets spoken by Sadat, Mubarak and their international backers notwithstanding, economic 'reform' above all reproduced authoritarianism and oligopolies (Kienle, 2001). The economy in which military officers, security hawks and neoliberal industrialists operated, with its markets dominated by cartels, for-sale licences and politically awarded contracts, generated spectacular aggregate statistics but no development. The post-1992 land tenure changes, euphemistically termed 'agricultural liberalization', restored the push for export agriculture and with it much of the power of landlords, who were able to raise rents dramatically after Law 96 was passed (Saad, 1999).

There was plenty of capital accumulation and financial speculation, but little income growth for the population, productivity increases or food security. As one World Bank study found: "Household welfare in real terms has not improved overall between 2000 and 2009 and has declined for most households.... The gap between GDP per capita and household consumption has increased during the last decade. While GDP per capita has grown steadily, household consumption has not increased" (Verme, 2014, p. 95). Another analysis highlighted that, in parallel with rapid growth and ameliorating Egyptian competitiveness (International Monetary Fund, 2010), "income poverty increased from 19.6% in 2004/2005, to 21.6% in 2008/2009, to 25.2% (21 million people) in 2010/2011.... Chronic malnutrition among children started to rise as early as 2003, and by 2008 about one-third of Egyptian children under the age of five were stunted" (Joint IFPRI-WFP Country Note, 2013). Massive quantities of American aid in the form of wheat and large-scale (commercial) cereal imports from across the world are the flip side of failed agricultural policies that ended up subsidizing urban consumption rather than strengthening rural livelihoods. The political correlate of this exclusionary growth model was mounting social unrest and violent protests in the countryside that pre-dated the 2011 Egyptian Revolution (Bush, 2011).

With unprecedented amounts of capital circulating as a result of primitive accumulation, foreign aid flows and remittances from the Egyptian diaspora, rent-seeking in the partly liberalized Sadat-Mubarak economy triggered a real estate boom, but not an investment splurge where it was really needed. The political exigencies of elite competition produced economic and environmental policies that focused public funds and laws almost exclusively on the supply side of the food-and-water equation but did little productive on the demand side. Whereas the latter would almost certainly entail the economic and political empowerment of disenfranchised population groups, the former were privileged because they could be framed in apolitical technospeak and dovetailed with the regime's logic of clientage. Pricey infrastructure ventures were undertaken under the banner of food self-sufficiency, an oft-repeated fantasy, disconnected from the reality of rising imports but congruent with the supply-side mantra. These projects really represented asset transfers of public funds to private hands, as with the infamous Toshka scheme, yet another Tahrir-style attempt to reclaim millions of *feddan* from the desert. Not coincidentally, opposition to such fantasies of greening the drylands is typically dismissed via the familiar refrain of warnings of a Malthusian collapse. Consider this backgrounder by the business advisory firm Kable Intelligence Limited:

> Over the past 20 years, the population of Egypt has risen from 20 to nearly 70 million and it has been predicted that this trend will continue, reaching an anticipated 120 million in the next 20.... Increasing urbanisation of the population places increased demands on the water supply, further exacerbating the problem for a country which is 95% desert. The Toshka project – an ambitious project to create a second Nile Valley, redirecting 10% of the country's allotment of water from the Nile via a massive irrigation scheme – arose as part of a plan to increase the inhabitable land from 5% to 25%. Toshka and the Southern Egypt Development Project aims to develop and extend agricultural production and create new jobs and population centres away from the narrow confines of the Nile Valley.... It has been called "Egypt's hope for the twenty-first century". (*Toshka Project – Mubarak Pumping Station*, http://www. water-technology.net/projects/mubarak/)

Yet the central idea behind these projects is not to feed Egypt's hungry masses or to raise farmers' incomes but to use billions of cubic metres of Nile water and the reclaimed lands to capitalize on rising global commodity prices and to strengthen regime allies. Landlords, powerful patrons in rural areas, were an important pillar throughout much of the decades-long rule of Mubarak's National Democratic Party (Arafat, 2009). Billions of dollars were pumped into Toshka, officially in collaboration with Gulf Arab and American investors, but with the bulk of risks accruing to the Egyptian treasury and the gains produced by these environmentally troublesome ventures flowing to Mubarak's crony capitalists and their foreign partners (Bush, 2007). Again, failure to meet the production or employment targets of propaganda bulletins is not the point. What matters is supercharging the sinews of financial power and political influence (Warner, 2013). The narrative of environmental determinism resonates with external observers who bemoan Egypt's multitude of social and environmental problems, but see no alternative to endorsing the autocrats responsible for those problems, for fear of state collapse and an advancing Islamist threat on the Nile (see e.g. Kaplan, 1996).

It is an iron, circular logic in which Egypt's geography and history are said to leave space for only one kind of political system and development model, rooted in a very particular framing of the links between climate, water supply and crop production. The resultant class structure and engagement with the global economic system then go on to cause policy stasis, environmental problems and increased social tensions (Richards, 1984); and the risk of a political explosion of these troubles is then invoked, again, to justify the very institutions, discourses and practices that gave rise to them in the first place.

Conclusion: from nexus, singular, to nexuses, plural

Nexus discussions in Egypt and elsewhere often occur in a vacuum, without reference to domestic political economies (and the ways in which they are imbricated in globalized networks and flows) or to political imaginaries. Yet nexus thinking is nothing new, with the Nile Basin as a poignant example: state-building projects and regime consolidation strategies there have for generations been premised on it. As I demonstrated elsewhere in the case of Sudan, the logic of water, civilization and power has been central, both as material reality and as political imaginary, in the creation of the modern Sudanese state, its recurrent state-building attempts, and the strategy of its elites to attain hegemony (Verhoeven, 2015). The latest terminology for interconnectivity of resources and scales – 'the nexus' – is seen in Egypt, Ethiopia and Sudan as yet another trope to be leveraged, at local, national and regional levels; it is an instrument for tilting tussles over power to one side or another. The malleability of nexus discourse allows it to be put at the service of very different but all resolutely political projects.

Mainstream sustainable development, to which the idea of 'the nexus' belongs, draws on the teleological assumption that bitter politics and associated rent-seeking in the water, energy and food sector are phenomena of the early stages of growth. Much nexus literature falls prey to a Rostow-like modernization paradigm, with the irresistible advance of technology and good governance steamrolling inefficiencies and irrational allocations. Yet the creation of a specific kind of political economy, with lucrative benefits for insiders and high costs for the multitude, forms the very core of the hydraulic missions emerging during industrial modernity. It is no coincidence that the pursuit of nexus reveries – the arrival of electricity in the peripheries, the conversion of peasant farmers into commercial farmers supplying international markets, the blooming of the desert – is closely linked to other modernizing dreams in the form of the expansion of the state and the awakening of the nation. From California and Mexico to South Africa, political elites and their bureaucratic allies have sought to usher in a new era through 'projects of the century' that highlight certain interconnections between water management, food production and energy security while obfuscating alternatives that might empower subaltern groups or rival elites (Swatuk, 2010; Worster, 1985). "Hydrocracies" – the institutions which build and manage the nexus between dams, large-scale irrigation schemes and electricity grids (Molle, Mollinga, & Wester, 2009) – develop powerful corporate interests that propel forward the national mission, whilst driving a bureaucratic imperialism and enmeshing themselves in local and international circuits of accumulation (Singh, 1997).

As documented here and by other scholars, political survival and rent-seeking across the Nile Basin and the Middle East more broadly have been intimately correlated with tropes of modernization, environmental determinism and engaging the global political economy (Davis & Burke, 2011). State-building and nation-building efforts have waxed and waned as regional actors put their resources at the disposal of international capital in the belief that tight links with the dominant centres of global capitalism will allow domestic elites to extract rents from the international system that help maintain political control. With Egypt as a case in point, this deeply rooted process of extraversion implies that the Nile Basin's place in the global political economy is not a one-way route of exploitation but a conscious reproduction of dependency by regional elites as they find it expedient to quash domestic rivals and expand authority on their territory through global rent-seeking structured around the region's resources – water, crops, energy and people (Verhoeven, 2015). As demonstrated through the dissection of elite rivalries and the water, food and land policies they gave birth to in Egypt, the confluence of hydrocratic ideology, power politics and sequential rounds of primitive accumulation has sustained authoritarian governance and economic dead ends, with ruinous consequences for vast swathes of the population in terms of water and food security.

If what today we term 'the nexus' functions as a political commodity, this implies rethinking the fiction that technocratic fixes can solve environmental problems (Sowers, 2013). The struggle for sustainability and environmental justice is simultaneously one against political oppression; as one scholar observed 20 years ago, "the expansion of political participation is a necessary condition for more efficient allocation of the region's scarcest resource, water" (Richards, 1993, p. 224). The refusal to consider alternative nexuses and thus to make Egypt's political economy more equitable by boosting incomes and productivity in rural areas has manifested itself in an increasingly vulnerable economy, mired in debt. Moreover, Egypt's dependence on food imports and the 'demilitarization' of the army during the Sadat and Mubarak regimes have gravely tarnished its geopolitical position and the ability of Cairo's ruling circles to pull together to maintain their pre-eminent regional role – a position that Egypt has essentially forfeited in the Middle East and Africa in the last 30 years. Seldom has a regional hegemon so absent-mindedly destroyed its own relative lead over possible rivals as its military power has been deliberately curtailed by its president, its economic base has been sapped to let rents accrue to incumbents, and the contradictions of its development strategy have rendered it dangerously dependent on importing food and virtual water.

Studying these processes, then, highlights the ability of elites to manage various forms of resistance, but not in linear, teleological ways. Certain nexus constructions make tactical sense and stabilize contradictory and potentially threatening political-economic trends, but strategically the long-term price may be high, including the loss of hydro-hegemony. Ultimately, this points to the importance of refusing to accept that contemporary water–energy–food–climate interconnections, whether local, regional or global, are merely to be identified, measured and leveraged for opportunities. Though the technocratic paradigm suggests that the nexus is what you make of it, this paper has underscored that it is vital to understand that there is no nexus (singular) but only multiple, socially constructed nexuses. Thus, one suggestion is to give a radically different meaning to what a 'nexus approach' should become: a conversation that is no longer blind to its own subjectivity but rather a fundamentally political debate about how to reorganize dryland societies and how to restructure their relationship with the outside world.

Disclosure statement

No potential conflict of interest was reported by the authors.

References

Abdel-Fadil, M. (1975). *Development, income distribution and social change in rural Egypt 1952–1970* University of Cambridge Department of Applied Economics Occasional Papers, no.45. Cambridge: Cambridge University Press.

Adams, R. (1986). *Development and social change in rural Egypt*. Syracuse: Syracuse University Press.

Allan, T. (2003). IWRM/IWRAM: A new sanctioned discourse? Occasional Paper 50, SOAS Water Issues Study Group School of Oriental and African Studies/King's College London.

Allan, T. (2011). *Virtual water*. London: IB Tauris.

Anderson, D., & Grove, R. (1987). The scramble for Eden. In D. Anderson & R. Grove (Eds.), *Conservation in Africa. peoples, policies and practices*. Cambridge: Cambridge University Press.

Arafat, A. D. (2009). *The Mubarak leadership and future of democracy in Egypt*. New York, NY: Palgrave MacMillan.

Asian Development Bank. (2013). *Thinking about water differently. Managing the water-food-energy nexus*. Manilla: ADB.

Ayubi, N. (1980). *Bureaucracy & politics in contemporary Egypt*. London: Published for the Middle East Centre, St. Antony's College by Ithaca Press.

Baker, R. (1978). *Egypt's uncertain revolution under Nasser and Sadat*. Cambridge, MA: Harvard University Press.

Bakker, K., & Morinville, C. (2013). The governance dimensions of water security: A review. *Philosophical Transactions of the Royal Society A, 371*, 2002, 30 September.

Bazilian, M., Rogner, H., Howells, M., Hermann, S., Arent, D., Gielen, D., ... Yumkella, K. K. (2011). Considering the energy, water and food nexus: Towards an integrated modelling approach. *Energy Policy, 39*, 7896–7906. doi:10.1016/j.enpol.2011.09.039.

Brown, N. (1990a). Brigands and state building: The invention of banditry in modern Egypt. *Comparative Studies in Society and History, 32*, 258–281. doi:10.1017/S0010417500016480.

Brown, N. (1990b). *Peasant politics in modern Egypt: The struggle against the state*. New Haven: Yale University Press.

Bush, R. (2007). Politics, power and poverty: Twenty years of agricultural reform and market liberalisation in Egypt. *Third World Quarterly, 28*, 1599–1615. doi:10.1080/01436590701637441.

Bush, R. (2011). Coalitions for dispossession and networks of resistance? Land, politics and Agrarian reform in Egypt. *British Journal of Middle Eastern Studies, 38*, 391–405. doi:10.1080/13530194.2011.621700.

Butzer, K. (1976). *Early hydraulic civilization in Egypt*. Chicago, IL: University of Chicago Press.

Conca, K. (2006). *Governing water. Contentious transnational politics and global institution building*. Cambridge, MA: MIT Press.

Cook, S. A. (2012). *The struggle for Egypt*. Oxford: Oxford University Press.

Cuno, K. (1992). *The Pasha's peasants: Land, society, and economy in lower Egypt, 1740–1858*. Cambridge: Cambridge University Press.

Davis, D., & Burke, E. (Eds.). (2011). *Ecology and history. Environmental imaginaries of the Middle East and North Africa*. Athens: Ohio University Press.

de Guerville, A. B. (1906). *New Egypt*. London: William Heinemann.

De Schutter, O. (2011). How not to think of land-grabbing: Three critiques of large-scale investments in farmland. *Journal of Peasant Studies, 38*, 249–279. doi:10.1080/03066150.2011.559008.

De Stefano, L., Edwards, P., de Silva, L., & Wolf, A. T. (2010). Tracking cooperation and conflict in international basins: Historic and recent trends. *Water Policy, 12*, 871–884. doi:10.2166/wp.2010.137.

Eyre, C. (1994). The water regime for orchards and plantations in pharaonic Egypt. *Journal of Egyptian Antiquity, 80*, 57–80.

Eyre, C. (1999). The village economy in pharaonic Egypt. In A. Bowman & E. Rogan (Eds.), *Agriculture in Egypt. From pharaonic to modern times*. Oxford: Proceedings of the British Academy.

Fahmy, K. (2002). *All the Pasha's men: Mohamed Ali and the making of modern Egypt*. Cairo: The American University in Cairo Press.

Faris, M., & Khan, M. H. (Eds.). (1993). *Sustainable agriculture in Egypt*. Boulder: Lynne Rienner.

Fleming, J. R. (2010). *Fixing the sky. Fixing the sky: The checkered history of weather and climate control*. New York, NY: Columbia University Press.

Gibson, C. (1999). *Politicians and poachers. The political economy of wildlife policy in Africa*. Cambridge: Cambridge University Press.

Haddad, E. A. (2005). Digging to India: Modernity, imperialism, and the Suez Canal. *Victorian Studies, 47*, 363–396. doi:10.2979/VIC.2005.47.3.363.

Hansen, B. (1991). *The political economy of poverty, equity and growth. Egypt and Turkey*. Oxford: Oxford University Press/World Bank Group.

Harb, I. (2003). The Egyptian military in politics: Disengagement or accommodation? *Middle East Journal, 57*, 269–290.

Hashim, A. S. (2011). The Egyptian military, part one: From the Ottomans through Sadat. *Middle East Policy, 18*, 63–78. doi:10.1111/j.1475-4967.2011.00498.x.

International Monetary Fund. (2010). *Arab Republic of Egypt: Country Report No. 10/94*. Washington, DC: IMF.

Jensen, K. M. (2013). Viewpoint – swimming against the current: Questioning development policy and practice. *Water Alternatives, 6*, 276–283.

Joint IFPRI-WFP Country Note. (2013). Tackling Egypt's rising food insecurity in a time of transition. Retrieved from http://www.ifpri.org/sites/default/files/publications/ifpriwfppn_egypt.pdf

Kandil, H. (2011). Revolt in Egypt. *New Left Review, 68*, 17–55.

Kandil, H. (2012). *Soldiers, spies and statesmen. Egypt's road to revolt*. London: Verso.

Kaplan, R. (1996). *The ends of the earth*. London: Papermac.

Kienle, E. (2001). *A grand delusion: Democracy and economic reform in Egypt*. London: I.B. Tauris.

Koohafkan, P., & Stewart, B. A. (2008). *Water and cereals in drylands*. Rome: FAO/Earthscan.

Lélé, S. M. (1991). Sustainable development: A critical review. *World Development, 19*, 607–621. doi:10.1016/0305-750X(91)90197-P.

Magd, Z. A. (2010). Rebellion in the time of cholera: Failed empire, unfinished nation in Egypt, 1840–1920. *Journal of World History, 21*, 691–719.

Migdal, J. (1988). *Strong societies and weak states*. Princeton: Princeton University Press.

Mitchell, T. (1988). *Colonising Egypt*. Berkeley: University of California Press.

Mitchell, T. (2002). *Rule of experts*. Berkeley: University of California Press.

Molle, F., Mollinga, P., & Wester, P. (2009). Hydraulic bureaucracies and the hydraulic mission: Flows of water, flows of power. *Water Alternatives, 2*, 328–349.

Mortimore, M. (2010). Adapting to drought in the Sahel: Lessons for climate change. *Wiley Interdisciplinary Reviews: Climate Change, 1*, 134–143. doi:10.1002/wcc.25.

North, D., Wallis, J. J., & Weingast, B. (2009). *Violence and social orders: A conceptual framework for interpreting recorded human history*. Cambridge: Cambridge University Press.

O'Brien, P. (1968). The long-term growth of agricultural production in Egypt, 1821–1962. In P. M. Holt (Ed.), *Political and social change in modern Egypt* (pp. 162–195). Oxford: Oxford University Press.

Osman, T. (2010). *Egypt on the brink*. New Haven: Yale University Press.

Owen, R. (1993). *The Middle East in the world economy, 1800–1914*. New York, NY: I.B. Tauris.

Patel, R. (2007). *Stuffed and starved. Markets, power and the hidden battle over the world's food system*. London: Portobello Books.

Ribot, J. (2014). Cause and response: Vulnerability and climate in the Anthropocene. *The Journal of Peasant Studies, 41*, 667–705. doi:10.1080/03066150.2014.894911.

Richards, A. (1982). *Egypt's agricultural development, 1800–1980*. Boulder: Westview Press.

Richards, A. (1984). Ten years of infitah: Class, rent, and policy stasis in Egypt. *The Journal of Development Studies, 20*, 323–338. doi:10.1080/00220388408421921.

Richards, A. (1993). Economic imperatives and political systems. *Middle East Journal, 47*(2), 217–227.

Richter, T. (2007). The political economy of regime maintenance in Egypt: Linking external resources and domestic legitimation. In O. Schlumberger (Ed.), *Debating Arab authoritarianism* (pp. 177–194). Stanford: Stanford University Press.

Saad, R. (1999). State, landlord, parliament and peasant: The story of the 1992 tenancy law in Egypt. In A. Bowman & E. Rogan (Eds.), *Agriculture in Egypt. From pharaonic to modern times.* Oxford: Proceedings of the British Academy.

Singh, S. (1997). *The taming of the waters. The political economy of large dams.* Oxford: Oxford University Press.

Sowers, J. (2013). *Environmental politics in Egypt.* Abingdon: Routledge.

Springborg, R. (1979). PMiddle Eastern Studiesatrimonialism and policy making in Egypt: Nasser and Sadat and the tenure policy for reclaimed lands. *Middle Eastern Studies, 15,* 49–69. doi:10. 1080/00263207908700395.

Springborg, R. (1982). *Family, power and politics in Egypt.* Philadelphia: University of Pennsylvania Press.

Swatuk, L. (2010). The state and water resources development through the lens of history: A South African case study. *Water Alternatives, 3,* 521–536.

United Nations World Water Assessment Programme. (2014). *UN world water development report 2014. Water and energy* (Vol. 1). Paris: UNESCO.

Van De Walle, N. (2001). *African economies and the politics of permanent crisis, 1979–1999.* Cambridge: Cambridge University Press.

Vatikiotis, P. J. (1985). *The history of Egypt.* Baltimore: Johns Hopkins University Press.

Verhoeven, H. (2011). Climate change, conflict and development in Sudan: Global neo-Malthusian narratives and local power struggles. *Development and Change, 42,* 679–707. doi:10.1111/j. 1467-7660.2011.01707.x.

Verhoeven, H. (2014). Gardens of Eden or hearts of darkness? The genealogy of discourses on environmental insecurity and climate wars in Africa. *Geopolitics, 19,* 784–805. doi:10.1080/ 14650045.2014.896794.

Verhoeven, H. (2015). *Water, civilisation and power in Sudan. The political economy of Military-Islamist state-building.* Cambridge: Cambridge University Press.

Verme, P. (2014). Facts and perceptions of inequality. In P. Verme, B. Milanovic, S. Al-Shawarby, S. El Tawila, M. Gadallah, & E. A. A. El-Majeed (Eds.), *Inside inequality in the Arab Republic of Egypt.* Washington, DC: World Bank.

Wade, R. (1982). The system of administrative and political corruption: Canal irrigation in South India. *The Journal of Development Studies, 18,* 287–328. doi:10.1080/00220388208421833.

Warner, J. (2013). The Toshka mirage in the Egyptian desert – river diversion as political diversion. *Environmental Science & Policy, 30,* 102–112. doi:10.1016/j.envsci.2012.10.021.

Waterbury, J. (1983). *The Egypt of Nasser and Sadat: The political economy of two regimes.* Princeton: Princeton University Press.

Watts, M. J. (2004). Antinomies of community: Some thoughts on geography, resources and empire. *Transactions of the Institute of British Geographers, 29,* 195–216. doi:10.1111/j.0020-2754. 2004.00125.x.

Waughray, D. (Ed.). (2011). *Water security: The water-food-energy-climate nexus.* Washington, DC: Island Press/World Economic Forum.

Weinbaum, M. G. (1985). Egypt's infitah and the politics of US economic assistance. *Middle Eastern Studies, 21,* 206–222. doi:10.1080/00263208508700624.

Worster, D. (1985). *Rivers of empire: Water, aridity, and the growth of the American West.* Oxford: Oxford University Press.

Nexus meets crisis: a review of conflict, natural resources and the humanitarian response in Darfur with reference to the water–energy–food nexus

Brendan Bromwich

Department of Geography, King's College London, UK

Darfur has been widely used as a case study by both those arguing for causality between environmental scarcity and war and those disputing it. This article challenges that approach by drawing on debates taking place within Darfur, reflecting on both the conflict and the humanitarian response. It argues that reviewing Darfur on its own terms makes a stronger basis to identify transferable lessons for interventions elsewhere. It considers water, food and energy, and finds that supporting governance is an essential theme for promoting economic recovery and laying a foundation for a well-managed water–energy–food nexus.

Introduction: challenging the 'Darfur case study'

The considerable public awareness of the devastating conflict in Darfur has made it an enduring reference point in international debate over natural resources and conflict. However, whilst the gravity of what has taken place in Darfur and the traction with the general public raise interest in Darfur as a case study in such discourse, the scarcity of primary research, the complexity of the conflict, and the remoteness of the region to most western researchers mean a degree of caution is required in assessing the way in which foreign academics and activists draw on Darfur for debate beyond the immediate context.

Popular interest in global environmental issues and political activism converged in the competing narratives of Darfur with ecological and political-based conflict at times presented as a zero-sum game. Ecological narratives of the conflict have been rightly decried as providing a distraction from the political fundamentals (Verhoeven, 2011), but such arguments have in turn often masked the need for detailed, contextualized analysis of natural resources and conflict, for all that these dynamics may be subordinate to, and at times manipulated by, higher-level political processes. Reviewing climate change and human security in Africa, Srinivasan and Watson (2013, p. 307) point out that the local complexities are given insufficient weight in policy matters. Drawing on Brown (2010a), they remind us that "Africans are not really the intended audience of the Post Kyoto debate, but they are part of the evidence being used to make it." If that is true for Africa and climate in general, it has a particular resonance with Darfur and conflict over natural resources. This poses difficult questions about how we can move beyond unexamined selective interpretation of case studies or settling for the assertion that high politics are pre-eminent, leaving local resource management and governance issues unexamined.

Darfur has a long history of conflict at numerous levels. Tribal militias contesting land rights on the margins of shifting national and global power struggles have been a recurring feature since the Ottoman era. Authors stressing the links between natural resources and conflict point to the burgeoning population and declining rainfall between the 1950s and 1980s ending in the massive upheaval of the drought in 1984 and 1985 (Sachs, 2006; UNEP, 2007). It is reasonable to argue that the societal changes as a result of both the long-term changes and the massive shock of the drought are important elements of the backdrop to the Arab-Fur war of 1987–89. However, the rising trend of rainfall from the 1980s through to the early 2000s, when violence broke out on a scale unseen since the nineteenth century, must likewise be cited in order to temper the case for environmental determinism (Bromwich, 2009a). The pattern is reflected in a rising trend of vegetation as evidenced by remote-sensing analysis of North and West Darfur (Brown, 2010b). Kevane and Gray (2008, p. 2) point out that over a 30-year period prior to the conflict Darfur rainfall exhibited a "flat trend though with high variability".

A more detailed reading would suggest that the upheaval of the drought did undermine governance arrangements needed to manage conflict over resources in subsequent decades, as evidenced by the rising tension over access to land in 2001–03. Notwithstanding this tension, the trigger (and the vast influx of arms) that transformed the scale of the conflict was different: it was political in nature, and the grievances were focused against the national government. Disaffected Darfuris had published details of their complaints relating to marginalisation in development expenditures and under-representation in power in May 2000 (Daly, 2007).[1] In Darfur government targets were being attacked in early 2003 in addition to the ongoing violence relating to tensions over land (Daly, 2007). The issue of Darfur's marginalization had been brought to a head, in part, by its exclusion from the renegotiation of power and wealth in Sudan as progress towards the 2005 Comprehensive Peace Agreement was bringing the southern civil war to a close. However, the situation was further complicated by Islamist disaffection with the ruling regime and escalating racialization of Darfuri politics (Cockett, 2010; Daly, 2007). The counter-insurgency in Darfur saw the deployment of tribal militia and government air strikes combined with a massive influx of arms (Cockett, 2010). The supply of arms to the militia was, of course, fuel to the fire in the escalating tribal tensions in which land was a key element. The conflation of these different elements of conflict, spread out over huge tracts of land, goes some way towards explaining the scale and horror of the war that ensued.

The salient point here is that different parts of the complex recent history of Darfur can be taken in isolation to support different perspectives in debates outside Darfur. Such an approach, however, fails either to inform Darfuri activists in their search for peace, or to make a reliable contribution to debate on the generic links between conflict and natural resources elsewhere.

This article argues that to understand the political ecology of Darfur the region needs to be considered on its own terms. Given the difficulty of discerning which of the divergent narratives of Darfur to draw on, one test that can be applied is the extent to which the analysis gained traction within Darfur rather than solely in policy fora and academic discourse outside Sudan. Consultations relating to natural resources, conflict and governance in Darfur took place at political and technical levels. At times these levels were combined, as in the discourse on water resource management, which is a challenge with both technical and political dimensions in Darfur. Examples of policy consultations relevant to natural resources include the Darfur livelihood workshops in 2007 (Young et al., 2007), the El Fasher Climate Conference 2010 (UN, 2010a), the Darfur International Water Conference 2010 (UN, MWR, AU, 2011) and most significantly the Darfur Joint

Assessment Missions (DJAMs) in 2006 and 2013 (Darfur Regional Authority and Government of Sudan, 2013). The DJAMs were formal planning and consultation processes triggered by the signing of the Darfur Peace Agreement (2006) and the Doha Document for Peace in Darfur (2011), undertaken to prepare for major reconstruction programmes. In each case, the peace process has failed to bring an end to the multiple patterns of conflict, and therefore large-scale reconstruction has not taken place.

Notwithstanding the challenges of security and access, and the background of chronic under-funding of Darfur's universities, some important and at times impressive pieces of primary research undertaken in Darfur exist. A notable example would be Young et al.'s *Darfur: Livelihoods under Siege* (2005), informing the consultations in 2007 (Young et al., 2007) triggered by the Darfur Peace Agreement in 2006. One outcome of these consultations was a research theme on pastoralist livelihoods. The subsequent report (Young, Osman, Abusin, Asher, & Egemi, 2009) informed the 2010 El Fasher Climate Change workshop. The climate workshop was the basis of the UN report *Beyond Emergency Relief: Longer-Term Trends and Priorities for UN Agencies in Darfur* (2010b). Alongside this work led by Tufts University focusing on livelihoods, a series of locally produced research projects were undertaken, begun by Tearfund (2007a, 2007b) and continued by UNEP (2008a, 2008b), informing the same consultations, with a focus on water and environment. This program of work also informed the production of the integrated water resources management (IWRM) vision statements made by the Darfuri water sector in 2010 (MWR and UNEP, 2010a, 2010b), which were endorsed by federal government and reflected in the major Darfur water conference in 2011 (UN, MWR, AU, 2011). These programmes are ongoing.

This article draws on documents such as these, written primarily for an audience of Darfuri and international decision makers, peacemakers, environmentalists, humanitarian actors and development workers. The article aims to review the water–energy–food nexus in Darfur in its own context, and then draw out implications for management of the nexus in the Middle East and North African region.

Outline

Following on from this introduction, the second section provides an overview of the natural resources context of Darfur and considers how the geography is reflected in the different livelihoods pursued in the region. Livelihoods are important to this consideration of the nexus, as different groups use resources in different ways. This provides a basis to understand tensions between groups and the challenges of governance with respect to resources relevant to the nexus. The pattern of peace agreements relating to the region provides a framework for the various conflicts, enabling causal links to be traced with greater clarity.

The third section introduces impacts of both the conflict and the humanitarian response with regard to water, energy and food, in turn. The 'water–energy–food nexus' is a term used to draw attention to the interconnectedness of these three themes, the management of which may require trade-offs, prioritization or at least the search for coordination and complementarity. This discussion is expanded with a consideration of how the secession of South Sudan and the impact of the loss of oil in the economy have complicated the challenges posed for the nexus in Darfur.

This review of water, energy and food sets up the main discussion in the article, which develops the central theme of governance which links conflict mitigation and sustainable management of the nexus. The article concludes, firstly, with a call for complementarity between short-term humanitarian responses and longer-term promotion of a well-managed natural resource base as the foundation for economic recovery with an efficient water–energy–food nexus. Secondly, a call is made, in light of the experience of Darfur, to enhance the governance of natural resources in other conflict-prone states, because should conflict occur it would undermine the environmental governance essential to mitigate risks of further conflict.

The context and conflict in Darfur

The geographical context

Darfur is located in the Sahelian region in Western Sudan, between 11° and 17° north. The population of 8.5 million live almost entirely to the south of 15° due to the extreme aridity to the north. Isohyets for the region decrease from 800 mm in the south to 50 mm or less in the north, interrupted by the presence of the Jebel Marra massif, the centre of the region, with an elevation of 3088 m, acting as a central water tower. The drainage pattern of Darfur is made up of a number of major ephemeral rivers or 'wadis' draining radially out from Jebel Marra to the Chari Basin in Chad and the Central African Republic, to the Bahr El Arab and into the Sudd, and to the Nile in Northern State. In a number of places, the wadis (such as Wadi El Ku and Wadi Howar) do not reach the main water courses, but the water drains into aquifers or evaporates.

Of particular significance is the variability between seasons and from one year to the next. Figure A1 (in the online supplemental data, http://dx.doi.org/10.1080/07900627. 2015.1030495) shows the coefficient of variability of the annual maximum vegetation (Normalized Difference Vegetation Index). This provides a useful visualization of the Sahel (a word that means 'the shore' in Arabic), in which well and consistently vegetated Sub-Saharan Africa lies to the south, and the hyper-arid Sahara Desert to the north. The Sahel may be understood as a region of great variability lying between these two.

The Nubian sandstone to the north is an important aquifer with large volumes of fossil water. The Baggara Basin lies to the south and receives recharge from Darfur's south-flowing wadis. Due to their distance from major sources of population, these have not been developed commercially. Some smaller sandstone aquifers exist and are important for the water supply of El Fasher and El Geneina. Much of Jebel Marra is made of volcanic rocks, whereas much of the flatter, more populated land around the mountains is underlain by fractured basement complex rocks. The yield of boreholes in these rocks is dependent on their being connected to a network of fractures with sufficient recharge and storage capacity. These boreholes have low and variable yields (Tearfund, 2007b).

Some of the most productive wells lie in the water-rich alluvial deposits along the wadis. These alluvial aquifers are also important in feeding the adjacent and underlying aquifers in fractured basement complex rocks. Therefore, groundwater (as well as surface water) is generally more plentiful closer to wadis. This pattern of the availability of water is important in explaining the spatial variability of Darfur, which is largely made up of fertile water-rich wadis used for cropping and human settlements and drier interfluvial rangelands used less intensively for crops and more for migratory livestock production, supporting lower population densities.

Livelihoods, pastoralism, agro-pastoralism, agriculture trade

The livelihood patterns of Darfur reflect the spatial and temporal variability of the physical geography of the region. In addition to the richer cropping along the wadi valleys, localized patterns of agro-pastoralism are practised, within which livestock are moved on a relatively short-range migration between the wadis and the interfluvial rangelands. Longer-range pastoral migrations also occur, with north–south movements across the region. Camel-herding groups travel from the arid north as far as South Sudan and the Central African Republic on six main routes (Young et al., 2009). Cattle migrations are mainly in the central and southern latitudes. Whilst this picture creates an initial starting point for understanding the interactions of different livelihood groups in Darfur, the additional variables of soil type and elevation are important in determining the wider pattern of livelihoods and agricultural production. The main livelihood zones are shown in Figure A2 (in the online supplemental data).

The rationale for migration is that it is an effective response to the considerable variation in availability of resources in drylands. Water is a limiting factor on vegetation, and growth occurs rapidly following rains. By migrating, herds have the opportunity to feed on the grass at the most nutritious period of the cycle of growth, notwithstanding the localized and short-lived nature of rainfall. This system is therefore highly dependent on green water.[2] Regular access to sources of blue water is needed for stock watering and for the people accompanying the herds.

The variation in timing of rainfall is also a considerable challenge to farming communities. Analysis undertaken by UNEP, shown in Figure 1, indicates that the variability of the timing of rainfall and hence the duration of the cropping season has been a critical factor in the high incidence of crop failures in the El Fasher area. The impact of this variability could be mitigated to some extent with investment in rainwater-harvesting infrastructure, but high levels of poverty and limited government capacity to support uptake of such approaches increase the exposure of communities to these effects.

Change and governance: long-term environmental and governance changes

Population growth has had an important function in the expansion of agriculture in Darfur. Figure A3 (in the online supplemental data) shows changes in land use in a $100 \, km^2$ area in

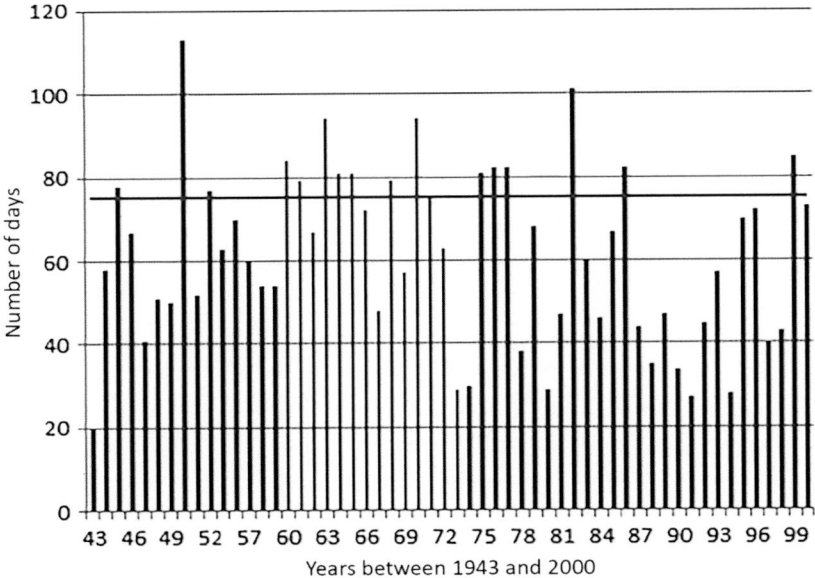

Figure 1. Length of the cropping season, 1943–2000, in the El Fasher area. The heavy horizontal line indicates the minimal length required for a good crop of millet (75 days). Source: UNEP (2014a, p. 10).

the Um Chaloota region of central Darfur between 1973 and 2000. The map has two lighter shades, pale green and pink, that indicate rangeland and rainfed agriculture. The two darker shades, green and light teal, represent forest and bushland. The most striking change in land use is the increase in rainfed agriculture at the expense of rangeland and forest. An important impact of these changes is on pastoralists, who in 1973 would have had easy access from the rangelands, through the cultivated area, to the wadis to water their herds, and for longer-distance migrations if required. In 2000, negotiating both long-distance migration and access to the wadi was much more demanding. This brings a focus onto governance arrangements for natural resources. Significantly more complex negotiation would have been needed in 2000 than in 1973 (Bromwich, 2009a).

Conflict

The opening discussion of this article highlighted the pitfalls of attaching narratives to elements of the Darfur conflict selectively. Mindful of these risks, a means of organizing references to the conflict is still needed that strikes a balance between oversimplification and trepidation at the complexity. A step towards striking this balance can be made for the Darfur conflict with a framework of three interconnected levels: local, national and international (Young et al., 2005). The justification for this selection, and identification of the issues at stake at each level, is made by examining the peace processes, which are themselves organized with these geographical scopes (UNEP, 2014b, 2014c).

The *international* regional power struggles involving Sudan, South Sudan, Uganda, Central African Republic, Chad and Libya have a history of many decades. A Chadian rebel movement used Darfur as its base to launch an attack on N'Djamena in February 2008 (a pattern echoing regional conflicts both in the 1960s and in the 1980s), confirming an international dimension of the tensions in Darfur at this time (Giroux, Lanz, & Sguaitamatti, 2009). Identity and regional power have been important dynamics of these international struggles.

At the *national* level, violence relating to the Darfur conflict has also spilled over beyond the geographical boundaries of Darfur. In May 2008 a Darfur rebel movement attacked greater Khartoum in central Sudan. The drivers for the conflict at the national level can be seen in the contents of the peace treaties negotiated to address it. Both the 2007 Darfur Peace Agreement and the 2011 Doha Document for Peace in Darfur have core chapters on wealth sharing and power sharing, indicating the fundamental political issues at stake. Rebel groups have made alliances within the country beyond Darfur, notably now linking with groups fighting in Blue Nile State and South Kordofan.

At the *local* level, conflict often occurs between communities self-identified along tribal lines, and both anecdotal evidence and the content of local-level agreements indicate that access to and control of natural resources, particularly land, is an important factor in these conflicts (UNEP, 2014c).

The Doha Document for Peace in Darfur (2011, p. 39) implicitly accepts that there have been inadequate governance arrangements to manage competition over resources at the local level when it states that "competition over pasture and water between herders and farmers is a serious problem" and calls for new policies and a "framework for equitable access for various users of land and water resources". Many commentators on Darfur endorse this analysis, citing failures of governance as a critical factor in local conflict (Abdul-Jalil, Mohammed, & Yousef, 2007; De Waal, 2007; Young et al., 2005).

Whilst the conflict broadly operates along these levels, one of the major difficulties in the overarching search for a political solution is how different levels of the conflict interact, as evidenced above relating to the counter-insurgency of 2003–05. The use of tribal militia in conflicts in wider disputes, notably between central elites and underdeveloped peripheral regions, has a long and bloody history in Sudan. This indicates that both enhanced governance of land and natural resources *and* national political reconciliation are prerequisites for lasting peace in Sudan.

Discussion

A number of observations on the Darfur context can be made based on this introduction, providing a platform for the discussion of the nexus that follows. Firstly, the humanitarian and peace-building response to the conflict has taken place in an ecological area characterized by scarcity and variability in resources. This presented a new challenge to humanitarians to integrate environmental concerns into their response (Tearfund, 2007a; UNEP, 2007). Whereas responses in areas such as South Sudan, Sierra Leone, Cote d'Ivoire, Democratic Republic of Congo, etc. had faced similar challenges of managing camps for displaced people in the context of insecurity, Darfur required a mix of skills – those used to working in drought responses needed to merge approaches with those experienced in addressing conflict.[3]

Secondly, natural resource variability and livelihoods are inextricably linked in Darfur. There are attributes of traditional management of livelihoods that have adapted to this variability, such as the reliance on migration, which need to inform management of the components of the nexus. At the same time, there is an imperative to modernize and intensify production.

Thirdly, governance may be understood as facilitating the interaction of different resource users. It is therefore an important enabler in adaptation of different natural resource–based livelihoods. Long-term changes are stretching governance arrangements in Darfur and are rightly seen as a contributing factor in the conflict. However, conflict (at all three levels) has also undermined governance, not least environmental governance (Bromwich, 2009b; UNEP, 2014b). Therefore, linkages between conflict and weak governance need to be understood as going both ways.

These considerations indicate how important it is that the humanitarian program in Darfur respond to the particular demands of the context in enabling Darfuris to rebuild and adapt environmental governance as part of Darfur's efforts to find lasting peace and to manage the challenges posed by the nexus in this context. The next section will introduce the different elements of the nexus, with consideration of the impacts of the conflict and of the humanitarian response, mindful of the centrality of environmental governance both to management of the nexus and to enabling Darfur to escape the recurring cycles of conflict described above.

Introducing the nexus

Water

Conflict has caused a loss of both natural resources and physical assets, combined with failures in both customary and formal governance arrangements. The major impact of conflict in the water sector is the massive concentration of demand in cities. Nyala, for example, grew from around 400,000 people to perhaps three times that many. This city lies on basement complex geology and faces extreme shortages during the dry season, with widespread drying of wells. Studies indicate that across Darfur some 800,000 internally displaced persons (IDPs) out of over 2 million may be at risk of significant groundwater depletion (Tearfund, 2007b, p. vii; UNEP, 2008a, p. 19). At two camps, alternatives to groundwater have had to be established following acute groundwater depletion. In addition to lost natural resources, assets such as water points have been destroyed as part of the conflict.

Conflict has undermined collaboration between livelihood groups in rural areas, creating difficulties for pastoralists and displaced communities in accessing water points. IDPs are not safe travelling in many locations outside the camps; in other locations it is the pastoralists who are not safe. These tensions were building up heavily in the years preceding the main outbreak of violence. Government has not been able to access many areas, and so services have been lacking. In addition, the fact that the government is seen as a party to the conflict in many areas has hindered it from performing a mediating role.

An impressive achievement has been made in providing lifesaving water, sanitation and hygiene (WASH) services to over 2 million people for a decade in extremely difficult operating conditions. UNICEF and the government Water and Environmental Sanitation Project have led that program to great effect (PWC UNICEF, 2011; UN, MWR, AU, 2011). One difficulty arises, however, with respect to the exit strategy. Humanitarian work is rights-based and draws on Sphere standards for humanitarian programming, under which the volume of water to be provided is widely interpreted as being 15 litres per person per day (Sphere Project, 2011). Anecdotal reports indicate that this may be more than rural Darfuris are accustomed to using, leading to some distortions in aid provision and a disincentivization of return. In the Abu Shouk camp, Oxfam found that only around 50% of the clean water provided was used for domestic purposes; otherwise, it was sold to neighbouring El Fasher or used for brickmaking (Oxfam, 2007). Groundwater in Abu Shouk has faced severe depletion, with water levels dropping by 7–10 m. More recently, a pipeline has been constructed to provide water to the camp. The challenge this lays down for those supporting Darfur's water sector is how to increase expectations of access to clean water and yet to enable communities to live within the limits of the scarcity and variability of Darfur's water resources.

As a means of re-establishing and rebuilding governance arrangements, a shared vision and practical efforts in establishing IWRM approaches in Darfur have been made. An ongoing exchange program with South Africa was a critical factor. South African efforts to transform the water sector into an ethnically inclusive model rather than one captured by a politically dominant elite as under apartheid resonated with Darfuri stakeholders. Building on the vision statements of 2010 (MWR and UNEP 2010a, 2010b), work is now being undertaken in one of Darfur's larger catchments, exploring the contribution of IWRM to rebuilding governance alongside the provision of practical assistance to conflict-affected communities (UNEP, 2014a).

Food

The conflict has had a considerable impact in undermining food production, as a result of displacement, asset stripping and insecurity. Over 2 million people were displaced and therefore ceased their farming and herding activities. There was widespread looting of assets, not least of cattle from the traditional agro-pastoral communities such as the Fur and Masaleet.

The pattern of livelihood activities began to change, including unsustainable, short-term coping strategies. Brickmaking was often taken up as a result of the high demand and because it could be undertaken close to the IDP camps. This resulted in a high degree of erosion of the topsoil in the rich alluvial areas close to the towns, undermining longer-term livelihoods and local food security. There was an increase in 'maladaptive livelihoods', defined as those with a negative impact either on their own sustainability or on the livelihoods of others.

Two examples of the types of changes taking place can be seen amongst northern camel-herding Abballa groups. The long-range herding of camels had been severely undermined by insecurity and the closure of the border for export to Libya. Various groups found important substitutes to their income through joining salaried militia or through shifting into firewood collection – an activity previously undertaken by the now-displaced communities (Young et al., 2009).

Transport of goods to markets was severely undermined, notably by the increase in travel times. Livestock taken on the hoof from Geneina to Omdurman would take 45–60 days prior to the conflict, but over 4 months during the conflict (Buchanan Smith, Fadul, Tahir, & Aklilu, 2012).

As of September 2012, 2.03 million people in Darfur were dependent on food aid, of whom 1.39 million were in IDP camps (UN, 2013, p. 18). Displacement is the largest cause of food insecurity, but the influence of the conflict goes far beyond those who are displaced. The World Food Programme's assessment of the food security of different livelihood groups is shown in Figure 2.

The importance of environmental governance is again highlighted in the examples here of how livelihoods have been undermined. This is accentuated by the fact that in the future there will be a need to include displaced populations returning to the rural communities, and to enable an increase in food production in order to end the reliance on food aid.

Energy

Darfuri communities are to a very large degree dependent on woodfuel for energy. Augmenting this, electricity in the main cities is provided from diesel generators,

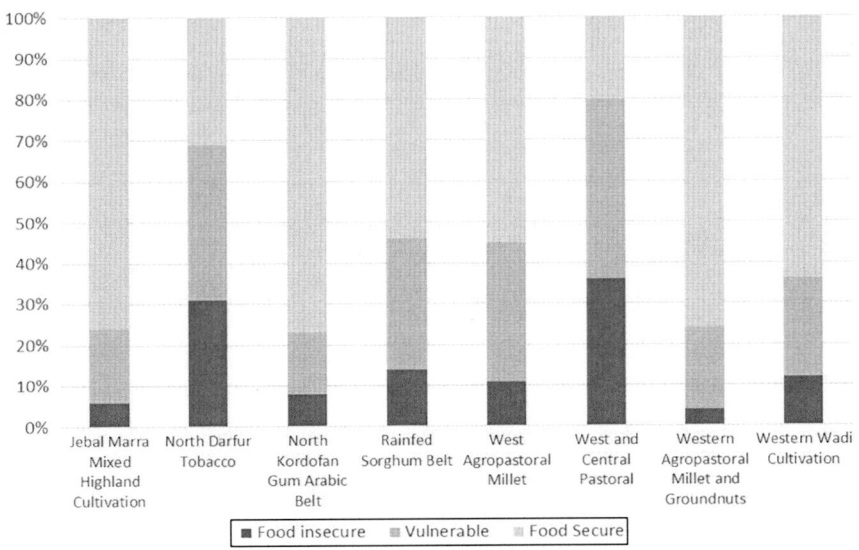

Figure 2. Food insecurity by livelihood zone in Darfur, 2012–13. *Source*: WFP (2014, p. 24).

and liquid petroleum gas (LPG) is used alongside woodfuel for cooking in many urban areas.

The FAO's Woodfuel Integrated Supply Demand Overview Mapping (WISDOM) analysis, shown in Figure A4 (in the online supplemental data), provides an indication of the environmental footprints of urban populations with respect to woodfuel (FAO, 2005, 2011). This bears out empirical observations that there are areas of significant deficit and other areas of surplus, creating a large volume of long-distance trade in woodfuel within both Darfur and Sudan more broadly (UNEP, 2007, 2008b). It emerges as an important livelihood in remote areas. It is therefore not surprising that pastoralist communities have taken the opportunity to move into wood collection in response to the contraction of their own economic activities and the displacement of the previous custodians of woodlands. A concurrent change taking place in rural Darfur was the breakdown of traditional governance arrangements for forestry (UNEP, 2008b; Young et al., 2009). In some cases trees were destroyed as acts of war (De Waal, 2004; Tearfund, 2007a), but as the economic importance of woodfuel increased, the patterns of collection became the focus.

A study undertaken by UNEP in 2008 to review the impact of conflict on the timber and woodfuel trade identified a massive increase in demand for wood for fuel and construction in response to the booming property market. For example, in Geneina, capital of west Darfur, the changes outlined in Table 1 between the pre-conflict situation and 2008 indicate this trend. The table shows a significant increase in rental prices, particularly for the higher-quality properties. The nine-fold increase in brickmaking indicates a massive response to this increased demand for property. At the same time livestock, one of the traditional areas for investment, was becoming less attractive due to widespread looting of cattle associated with the conflict (UNEP, 2008b).

Demand for woodfuel was also increasing in the urban areas, creating an important livelihood opportunity for displaced people in the camps around the cities. By 2006 NGOs had realized that women were putting themselves at immense risk not simply to collect firewood for cooking their food rations but also to sell woodfuel for cash for other objectives – i.e. they were pursuing a livelihood (Women's Commission for Refugee Women and Children, 2006). Whilst the demand for bricks was clearly a driver for the growth evidenced by the figures in Table 1, the increase in enterprises such as sawmills and carpentry shops also needs to be understood with the collapse of many other livelihoods as a result of insecurity. Many people were seeking to join one of the few areas of growth driven by the aid economy (UNEP, 2008b).

What emerges from this is the importance of understanding the economic distortions caused by both conflict and the international humanitarian and peacebuilding responses, including the influences these have on livelihoods. Livelihoods are crucial in that they are

Table 1. Changes in the housing and furnishing markets impacting timber demand in Geneina, West Darfur, pre-conflict and 2008.

	Pre-conflict	2008
Rent (first-class housing)	SDG 150–200	SDG 2000–3000
Rent (second-class housing)	SDG 75–100	SDG 400–500
Number of bricks made (from tax records)	9.0 M	82.6 M
Number of sawmills	10	17
Number of carpentry shops	65	120

Source: UNEP (2008b, pp. 15, 16, 26).

fundamental to the decision-making rationales of individuals and communities (UNEP, 2014c).

Deforestation in Darfur has been more intense closer to the cities, as the WISDOM mapping indicates. This has led to severe environmental degradation, often in areas which are important for agriculture. In addition to the deforestation that has already occurred, there is potentially a high demand for timber should any large-scale returns take place and villages be rebuilt. Some significant efforts have been made to promote the uptake of alternative construction technologies within the humanitarian response (UN-Habitat, 2012). This models good practice in terms of providing short-term emergency response (such as rebuilding schools and clinics) with long-term benefits that enable a well-managed nexus (capacity to build in more sustainable ways). Overall, however, the efforts have not as yet developed a local market for new technologies that has become self-sustaining.

Economic changes and the secession of South Sudan

This discussion has focused on the implications of the conflict within the boundaries of Darfur for management of food, water and energy. One of the major impacts from wider conflict dynamics has come as a result of the secession of South Sudan in 2011 and the consequent loss of oil revenue to Sudan. The Sudanese economy has experienced a significant contraction, with the value of the Sudanese pound on the black market weakening from 2.7 to 9.5 to the dollar. The impact on the price of food has been severe. The World Food Programme has tracked a steep rise in the cost of a 'local food basket' in Darfur (Figure 3).[4]

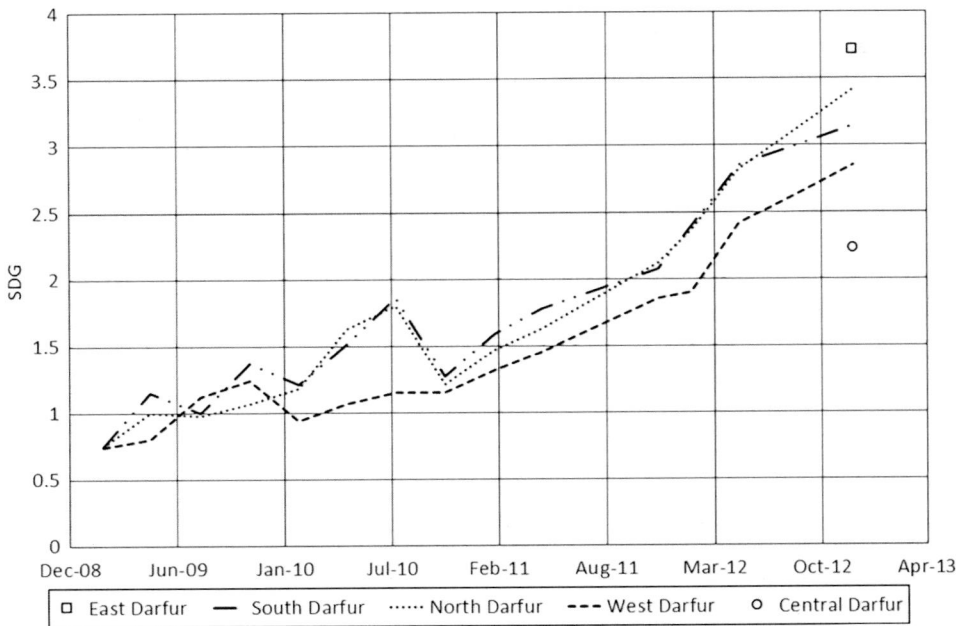

Figure 3.   Costs of a standardized 'local food basket' in Darfur, February 2009 to December 2012.
Source: WFP (2014, p. 18).

This demonstrates the need for a broader consideration of the nexus in Sudan's political economy, beyond the situation in Darfur. Key factors to consider include the impact of national-level fiscal and balance-of-payments pressures. It was the threatened reduction of fuel subsidies in Sudan that brought unrest to the streets of cities in central and eastern parts of the country in October 2013.

The knock-on impact of fuel prices on goods that are transported long distances is considerable, particularly in Darfur. Work on LPG undertaken by UNEP in 2010 identified that the cost of a 12.5 kg bottle of LPG was SDG 16 in Kosti in central Sudan but SDG 30 or SDG 35 in El Fasher and as much as SDG 50 in Geneina or Kutum (UNEP, 2010). The impact of transport costs bears out the adage that Darfur sells at the lowest prices and buys at the highest prices. In remote areas with poor infrastructure, the impact of energy prices on food will need particular attention.

Governance, conflict and the humanitarian response

This article has demonstrated the centrality of governance as a theme for both conflict mitigation and management of the nexus in Darfur. On this basis two important questions arise. What form of governance will enable sustainable and equitable management of the nexus in the post-conflict context? And, how can the humanitarian and peace-building work undertaken at this stage contribute to the development of suitable governance arrangements for this objective?

Equitable environmental governance

In reviewing governance, a comparison of the relative roles of 'vertical' and 'horizontal' relationships can be made. In the context of service provision, the vertical relationship prevails: Citizens pay taxes and the government or an agent of the government provides services. In more traditional tribal society, there is an important horizontal element with respect to agreements between groups over control of natural resources (all the structure within the tribes is hierarchical). This balance is an important issue for Darfur as it seeks to establish governance arrangements that enable disputes to be resolved without violent conflict (UNEP, 2013a). There are weaknesses that need to be overcome in both arrangements. As the example in Um Chaloota suggests, local and informal arrangements for resource sharing may not be compatible with increasingly crowded and congested resource allocation and use. However, top-down governance approaches also reach limits of capacity. This limitation is illustrated by the widespread silting of small and medium-size reservoirs across Sudan. Expectations of centralized government capacity for maintaining all such reservoirs across Sudan are unrealistic. The question arising here is more specifically what actions government can take to enable collaboration amongst resource users, as part of a strategic set of services supporting livelihoods and natural resources.

One means by which communities in Africa are seeking to the combine the benefits of local decision making over natural resources with formal legal frameworks is by developing 'co-management' regimes. Borrini-Feyerabend, Pimbert, Taghi Farvar, Kothari, and Renard (2007) draw attention to joint decision making and power sharing in co-management arrangements. In drylands, approaches to group ranching, community-based forestry, and watershed management are important expressions of co-management and are being used in a number of areas in Africa (UNEP, 2013a). Efforts to develop solutions for local governance of natural resources need to continue to innovate in drawing on both government and community capacities to address such difficulties. Supporting this kind of innovation is an appropriate agenda for local peace-building and humanitarian aid programming.

Humanitarian programming and governance

A number of challenges in the arrangements for humanitarian programming exist that militate against long-term strategic efforts for support to areas like Darfur in the context of a crisis. Humanitarian funding and planning cycles are generally only 12 months long, during which time the results of such efforts are unlikely to be evident. In addition, staff turnover makes contextually nuanced programming more difficult due to the challenge of retaining institutional memory in project offices.

Humanitarian delivery mechanisms have a strong focus on rights, which is of particular importance given the context of conflict and displacement in which rights have been denied. To some degree this models a more vertical form of governance – the relationship with the service provider being the focus rather than collaboration with other communities. A challenge arises in designing aid delivery arrangements that provide services for humanitarian assistance and yet draw on the traditional patterns of resilience and self-sufficiency of the affected communities. It is encouraging that the concept of resilience is gaining increased attention in Darfur's aid programming as a means of addressing these challenges. UNEP (2014b) draws attention to the importance of relationships amongst and between communities in promoting resilience, and the role of government in enabling these.

With respect to water, energy and food, it is evident that in this predominantly subsistence economy the notion of trade-offs across the nexus is less significant than in more complex economies. However, it is ironic that one of the points of interconnection is associated with the presence of the international community. The aid economy and the war economy drive the construction boom, which creates the market for bricks, often made with water provided for free in the humanitarian response. This industry has a significant impact on the depletion of groundwater resources, the loss of fertile farmland in safe locations needed for food production, and deforestation, affecting Darfur's energy supplies (UNEP, 2013a). These changes are influencing livelihoods which, as has been shown, are an important influence on local conflict dynamics. Understanding these linkages therefore emerges as an important humanitarian and peace-building priority.

In this case, the humanitarian service provision stands in contrast to traditional governance arrangements, in which strict rules around sustainable use of natural resources featured prominently. Therefore, in addition to the negative impact on the natural resources themselves, the approach delinks perceptions of entitlements to resource use from sustainable yields of those resources. Ultimately, these issues need resolving, although not in the short term by undermining genuine entitlements to humanitarian assistance.

Water resource management faces particular challenges for attention and funding in humanitarian contexts. It doesn't fit within the narrative of providing life-saving assistance to a definable number of 'beneficiaries'. The number of beneficiaries served by an intervention is a headline figure in seeking justification for funding, and therefore interventions such as the provision of new wells provide an attractive option for funding agencies looking for short-term and visible results. The failure of the wider humanitarian community to monitor groundwater resources for the first three years of the crisis, notwithstanding the massive and unprecedented demands on Darfur's low-yielding basement complex aquifers, indicate the scope of this problem (Tearfund, 2007b). Once water resources had begun to be addressed, an additional weakness occurred with respect to the frequent failure of newly constructed rainwater-harvesting structures and small dams. Addressing these gaps requires a broader approach to water than WASH programming alone. The 2011 Darfur International Water Conference achieved this by having complementary themes on WASH and IWRM, making for both a stronger humanitarian response and a significant contribution to the development of IWRM in Sudan – and contributing to the longer-term management of the nexus (UN, MWR, AU, 2011). This example in water has broader application: where interdepartmental and intersectoral coordination in government needs strengthening to meet the challenges of the humanitarian crisis, such enhancements in governance are needed in the longer term as the challenges of the nexus become more complex, with a greater degree of connectivity and trade-off.

In addition to addressing governance issues within the strategic objectives of aid programming, there is a need to manage the unintended impacts of the humanitarian responses. One of the encouraging developments in the humanitarian response in Darfur has been the uptake of an 'environmental marker' in the annual planning and funding process to demonstrate that a basic level of environmental mitigation has been undertaken. This marker has now been taken up globally in the UN humanitarian planning systems.

Livelihoods and traditional governance systems in Darfur have been built around the need to respond to the variability of resource availability. Urban settlements and 'temporary' settlements in camps (which may last for years or indeed decades) rely on patterns of more regular resource use and therefore drive modification of traditional patterns of governance and management of natural resources. The crisis is thus a driver of changes in resource use – and control of resources. The challenge humanitarians face in this context is to understand these changes and support adaptations in management of resources that enable long-term governance of natural resources – thereby providing a platform for equitable management of the water–energy–food nexus.

Politics and governance

The process of restoring peace in Darfur will include a resolution of the means by which natural resources are governed and managed in the region. This article has illustrated that the humanitarian response is relevant to governance of the nexus in unintended as well as in intentional ways. Greater awareness of rights and entitlements is a beneficial impact; however, the short-term horizons and de-emphasis of sustainability and local ownership are negative impacts. Some effort has been made to address this by exploring new forms of environmental governance, particularly the contributions that may be made by community-based natural resource management and IWRM.[5]

These practical demonstrations of new forms of governance make a contribution to the wider and longer-term effort to bring a more equitable framework for sharing natural resources – thereby contributing to the long-term political resolution of some of the fundamental grievances driving conflict at the local level. It is clear that progress is needed in both political and non-political contexts to support positive outcomes in rebuilding governance as part of the transition to peace. Work in apolitical contexts, such as the dialogues on livelihoods and on climate change and much of the analysis of IWRM, have enabled discussion highly relevant to longer-term political resolution of some of the underlying drivers of local conflict.

Such dialogues have demonstrated a convening power pertinent to some of the root causes of conflict. Tufts University engaged with actors of diverse political backgrounds in their projects to research pastoralist livelihoods (Young et al., 2009). As is evident from the discussion above, this topic is one of immense political significance in Darfur. By maintaining a scrupulous attention to neutrality and addressing the topic from a technical perspective, and through the level of trust built over a number of years producing high-quality research outputs, the work had considerable significance in informing both peace-building and humanitarian contexts. Likewise, UNEP's work on IWRM achieved a significance based on its inclusivity and technical approach rather than confronting the politics of natural resources head-on (UNEP, 2013b, 2014b).

Conclusions

The move towards developing complementarity in technical efforts to promote new forms of environmental governance and political resolution of conflict emerges as one of the approaches developed in the Darfur response that has potential to make contributions elsewhere. More effort is needed. Ultimately, practical local explorations in governance must have political endorsement to make a valid contribution to the resolution of conflict (UNEP, 2014b). That is not to say, however, that it is only the politics that counts. As with the analysis of conflict, so with the international response; both political and technical perspectives are needed.

This article has sought to contribute to a nuanced view of the links between the water–food–energy nexus, natural resources and conflict. Both conflict resolution and management of the nexus require a coordinated approach to governance of natural resources. This makes the case for integration of improved governance into responses to conflict as a basis for the long-term management of natural resources and of the water–energy–food nexus.

To the humanitarian, peace-building and development communities Darfur lays down a challenge to continue to develop integrated approaches in both the technical and political arenas that support the development of environmental governance, as part of Darfur's search for lasting peace. To practitioners and policy makers working on the nexus, particularly in drylands, the conclusion that Darfur's pattern of weak governance is both a contributory factor to, and a victim of, conflict makes a compelling case for enhancing environmental governance in vulnerable areas already at risk of conflict and struggling to manage the nexus.

Disclosure statement

No potential conflict of interest was reported by the author.

Supplemental data

Supplemental data for this article can be accessed at http://dx.doi.org/10.1080/07900627.2015.1030495.

Notes

1. The Justice and Equality Movement published their grievances in the "Black Book" at this time. The other main active rebel groups were the Sudan Liberation Movement/Army and the Darfur Liberation Front.
2. Molden (2007) provides the following definitions: "Green water refers to rainwater stored in soil or vegetation, which cannot be diverted to a different use. Blue water is surface and groundwater which can be stored and diverted for a specific purpose."

3. Some practices had been developed in Somalia and Northern Kenya, for example in environmental management at the Dadaab camp, but such activities were slow to be taken up in Darfur. This may be a function of the scale of the response in Darfur, in which practitioners with experience in drylands were a relative minority.

4. "The local food basket consists of eight food items that are typically consumed across Darfur: cereals (sorghum), milk, dry vegetables, cooking oil, goat meat, cow meat, onions and sugar. The quantity of each food item is calculated to meet the minimum requirements of 2100 kilocalories per person per day" (WFP, 2014, p. 17).

5. UNEP have operated in Darfur since 2007 in order to coordinate efforts to mitigate negative environmental impacts of humanitarian programming and to support Darfuri efforts to rebuild and adapt sustainable and equitable forms of environmental governance. For more information see www.unep.org/sudan.

References

Abdul-Jalil, M. A., Mohammed, A. A., & Yousef, A. A. (2007). Governance in Darfur: Past and future. In A. De Waal (Ed.), *War in Darfur and the search for peace*. Boston: Global Equity Initiative, Harvard University.

Borrini-Feyerabend, G., Pimbert, M., Taghi Farvar, M., Kothari, A., & Renard, Y. (2007). *Sharing power. A global guide to collaborative management of natural resources*. Colombo: Earthscan.

Bromwich, B. (2009a). Analysing resource constraints as one dimension of the conflict in Darfur. In *Environment and conflict in Africa: Reflections on Darfur*. Addis Ababa: UPEACE.

Bromwich, B. (2009b). Environmental impacts of conflict: The loss of governance and routes to recovery. In *Environment and conflict in Africa: Reflections on Darfur*. Addis Ababa: UPEACE.

Brown, O. (2010a). Campaigning rhetoric or bleak reality? Just how serious a security challenge is climate change for Africa? In Heinrich-Böll-Stiftung (Ed.), *Climate change resources migration - Securing Africa in an uncertain climate*. Cape Town: Heinrich Böll Foundation.

Brown, I. (2010b). Assessing eco-scarcity as a cause of the outbreak of conflict in Darfur: A remote sensing approach. *International Journal of Remote Sensing, 31*(10).

Buchanan Smith, M., Fadul, A., Tahir, A., & Aklilu, Y. (2012). *On the Hoof: Livestock trade in Darfur*. Boston: Feinstein International Center, Tufts University.

Cockett, R. (2010). *Sudan: Darfur and the failure of an African state*. Haven, CT: Yale University Press.

Darfur Peace Agreement (DPA). (2006). Abuja, Nigeria. http://peacemaker.un.org/sites/peacemaker.un.org/files/SD_050505_DarfurPeaceAgreement.pdf (last accessed 8 February 2015).

Darfur Regional Authority and Government of Sudan. (2013). Developing Darfur: A recovery and reconstruction strategy.

Daly, M. W. (2007). *Darfur's sorrow: a history of destruction and genocide*. Cambridge: Cambridge University Press.

De Waal, A. (2004). Counter-Insurgency on the cheap. *London Review of Books* [Online], *26*, 25–27, Available from http://www.lrb.co.uk/v26/n15/alex-de-waal/counter-insurgency-on-the-cheap [Accessed 18 November 2014].

De Waal, A. (Ed.). (2007). War in Darfur and the search for peace. Boston: Global Equity Initiative, Harvard University.

Doha Document for Peace in Darfur (DDPD). (2011). Doha, Qatar. http://unamid.unmissions.org/Portals/UNAMID/DDPD%20English.pdf (last accessed 8 February 2015).

FAO. (2005). WISDOM – East Africa. Woodfuel Integrated Supply/Demand Overview Mapping (WISDOM) Methodology. Spatial woodfuel production and consumption analysis of selected African countries. Prepared by for the FAO Forestry Department – Wood Energy. www.fao.org/docrep/009/j8227e/j8227e00.HTM (last accessed 7/6/14).

FAO. (2011). *WISDOM Darfur land cover mapping and wood energy analysis of Darfur's IDP regions summary report*. Rome: FAO.

Giroux, J., Lanz, D., & Sguaitamatti, D. (2009). The tormented triangle: The regionalisation of conflict in Sudan, Chad and the Central African Republic. Working paper 49; Crisis States Research Centre; London School of Economics, London.

Kevane, M., & Gray, L. (2008). Darfur: Rainfall and conflict. *Environmental Research Letters, 3*(3), 034006.

Molden, D. (Ed.). (2007). *Water for life, water for good: A comprehensive assessment of water management in agriculture.* Colombo: International Water Management Institute, Earthscan.

MWR and UNEP. (2010a). South Africa Integrated Water Resource Management (IWRM) Decision-Makers' study tour recommendations made at wrap up meeting Cape Town 6 November 2010. http://unep.org/disastersandconflicts/CountryOperations/Sudan/Consultations/tabid/102251/Default.aspx (last accessed 18 November 2014).

MWR and UNEP. (2010b). Darfur - South Africa IWRM technical study tour vision statement and recommendations, Cape Town 28 May 2010 vision statement. http://unep.org/disastersandconflicts/CountryOperations/Sudan/Consultations/tabid/102251/Default.aspx (last accessed 18 November 2014).

Oxfam. (2007). *Abu Shouk and Al Salaam water resource and water usage survey report.* El Fasher: Oxfam.

PWC UNICEF. (2011). *External programme evaluation: UNICEF assisted water, sanitation, and hygiene programme in Sudan (2002–2010).* Khartoum: North South Consultants Exchange.

Sachs, J. D. (2006). Ecology and political upheaval. *Scientific American, 295,* 37–37, New York10. 1038/scientificamerican0706-37

Sphere Project. (2011). Humanitarian charter and minimum standards in humanitarian response. http://www.spherehandbook.org/en/water-supply-standard-1-access-and-water-quantity/ (Last accessed 8 July 2014).

Srinivasan, S., & Watson, E. (2013). Climate change and human security in Africa. In R. Redclift & M. Grasso (Eds.), *Handbook on climate change and human security.* Cheltenham: Edward Elgar.

Tearfund. (2007a). *Darfur: Relief in a vulnerable environment.* Teddington: Tearfund.

Tearfund. (2007b). *Darfur: Water supply in a vulnerable environment.* Teddington: Tearfund.

UN. (2010a). Climate change: Retreat on adapting to climate change in the three Darfur states – Vision document. 23-24 March 2010, El Fasher. http://climatechange.sudanct.net/vision (last accessed 18 November 2014).

UN. (2010b). *Beyond emergency relief: Longer-term trends and priorities for UN agencies in Darfur.* Khartoum: United Nations.

UN. (2013). *SUDAN: United Nations and partners work plan 2013.* Khartoum: United Nations.

UN, MWR, AU. (2011). Darfur International Water Conference on water for sustainable peace. Donor appeal document, Khartoum.

UN-Habitat. (2012). *Economic benefits of stabilised soil block technology in Sudan.* Nairobi: UN-Habitat.

UNEP. (2007). *Sudan post conflict environmental assessment.* Nairobi: UNEP.

UNEP. (2008a). *Water resource management in humanitarian programming in Darfur: The case for drought preparedness.* Geneva: UNEP.

UNEP. (2008b). *Destitution, distortion and deforestation: The impact of conflict on the timber and woodfuel trade in Darfur.* Geneva: UNEP.

UNEP. (2010). *The use of Liquified Petroleum Gas (LPG) in Sudan.* Khartoum: UNEP.

UNEP. (2013a). *Governance for peace over natural resources: A review of transitions in environmental governance across Africa as a resource for peacebuilding and environmental management in Sudan.* Khartoum: UNEP.

UNEP. (2013b). *Republic of the Sudan: Country programme evaluation.* Nairobi: UNEP.

UNEP. (2014a). *Wadi El Ku catchment management project: Project inception report.* Khartoum: UNEP.

UNEP. (2014b). *Relationships and resources: Environmental governance for peacebuilding and resilient livelihoods in Sudan.* Nairobi: UNEP.

UNEP. (2014c). *Local level agreements in Darfur: A review with reference to access and management of natural resources.* Khartoum: UNEP.

USAID/FEWSNET. (2011). *Livelihoods zoning plus activity in Sudan aspecial report by the famine early warning systems network (FEWS NET).* Khartoum: USAID.

Verhoeven, H. (2011). Climate change, conflict and development in Sudan: Global neo-malthusian narratives and local power struggles. *Development and Change, Institute of Social Studies, The Hague, 42,* 679.

WFP. (2014). *Darfur comprehensive food security assessment SUDAN, 2012–2013.* Khartoum: World Food Programme. 2013.

Women's Commission for Refugee Women and Children. (2006). *Finding trees in the desert: Firewood collection and alternatives in Darfur.* New York: Women's Commission for Refugee Women and Children.

Young, H., Osman, A. M., Abusin, A. M., Asher, M., & Egemi, O. (2009). *Livelihoods, power and choice: The vulnerability of the Northern Rizaygat, Darfur, Sudan.* Boston: Feinstein International Center, Tufts University.

Young, H., Osman, A. M. K., Aklilu, Y., Dale, R., Badri, B., & Fuddle, A. J. A. (2005). *Darfur: Livelihoods under siege.* Medford, MA: Tufts Feinstein International Famine Center.

Young, H., Osman, A., Buchanan-Smith, M., Bromwich, B., Moore, K., & Ballou, S. (2007). Sharpening the Strategic Focus of Livelihoods Programming in the Darfur Region: A report of four livelihoods workshops in the Darfur region (June 30 to July 11, 2007). Khartoum: United Nations.

To what end? Drip irrigation and the water–energy–food nexus in Morocco

Guy Jobbins[a], Jack Kalpakian[b], Abdelouahid Chriyaa[c], Ahmed Legrouri[b] and El Houssine El Mzouri[c]

[a]Overseas Development Institute, London, United Kingdom; [b]Al Akhawayn University in Ifrane, Morocco; [c]Institut National de la Recherche Agronomique, Settat, Morocco

This article draws on three case studies of drip irrigation adoption in Morocco to consider the water–energy–food nexus concept from a bottom-up perspective. Findings indicate that small farmers' adoption of drip irrigation is conditional, that water and energy efficiency does not necessarily reduce overall consumption, and that adoption of drip irrigation (and policies supporting it) can create winners and losers. The article concludes that, although the water–energy–food WEF nexus concept may offer useful insights, its use in policy formulation should be tempered with caution. Technical options that appear beneficial at the conceptual level can have unintended consequences in practice, and policies focused on issues of scarcity and efficiency may exacerbate other dimensions of poverty and inequality.

Introduction: the water–energy–food nexus

It has long been recognized that water, energy and food are interdependent. The interconnections between these resources in planning and policy have been a long-standing issue explored in the literature (see e.g. Allan, 1997; Batliwala, 1982; Greeley, 1987; Keeney & Wood, 1977; Sachs, 1984). The water–energy–food (WEF) nexus has also been promoted by the international community: in 2011, the World Economic Forum and a conference held in Bonn both recognized the interdependence of water, energy and food security (Hoff, 2011; Waughray, 2011). Both argued that these should not be treated in separate silos by policy and planners, concluding that although problems are systemic, it is the world's poor who are most at risk from the scarcity and mismanagement of water, energy and food.

Central to the arguments of the WEF nexus is that increasing demand due to demographic and economic growth, coupled with stresses on supply resulting from factors such as climate change, are leading to shortfalls in availability of and access to food, energy and water (Waughray, 2011). Resource scarcity necessarily implies trade-offs between uses, such as choosing between allocating water to irrigation or to hydropower in times of drought. Conversely, investments to address one aspect of insecurity can exacerbate other aspects. For example, increasing water supply through desalination or groundwater pumping is highly energy-intensive (Siddiqi & Anadon, 2011; Trieb & Müller-Steinhagen, 2008). Similarly, exposure to risks in one dimension can exacerbate insecurity in the other dimensions, as demonstrated by the potential for increased energy prices to result in higher food prices (see e.g. Zhu, Ringler, & Cai, 2007). The interactions between these three kinds of security mean that attempting to address insecurity in a piecemeal fashion can result in a zero-sum game (Bizikova, Roy, Swanson, Venema, & McCandless, 2013).

WEF nexus approaches are therefore advocated as means of optimizing resource-allocation decisions and addressing unsustainable growth (Hoff, 2011; Waughray, 2011). A review of WEF nexus conceptual frameworks by Bizikova et al. (2013) concluded that their common goal is to promote action by identifying policy entry points to reduce trade-offs between types of security, improve all three types simultaneously, and exploit synergies. There are also concerns with, and connections to, other dimensions of human security and development, including international trade and investment (Allan, 2003; Allan, Keulertz, & Sojamo, 2012), ecosystem services (Rasul, 2014), and transitions to new economic systems such as the green economy (Hoff, 2011).

Despite this common goal noted by Bizikova et al., there are some differences in focus between these frameworks. For example, the World Economic Forum report emphasizes broad public policy areas such as trade, national security, finance, and business. It concentrates principally on macro-scale considerations of optimal resource allocation to avoid conflict and ensure economic growth (Waughray, 2011). By contrast, the Bonn2011 conference report is more aspirational, and identifies a series of options for economic transitions to promote inclusive and sustainable social and economic development (Hoff, 2011).

In the context of these different frameworks, the International Conference on Water-Food-Energy Nexus in Drylands, held in Rabat, Morocco, 11–13 June 2014, challenged participants to consider new ideas, how the WEF nexus might be moved from concept to application, and how it might benefit small farmers in global drylands. The present authors have responded to this challenge by drawing on three case studies from their research in Morocco that focus on the uptake of drip irrigation. We start by demonstrating that drip irrigation is an example of a nexus idea, and propose a simple framework to assess whether small farmers benefit from it. The article then outlines the key issues of water, energy and agriculture in Morocco and policies supporting the adoption of drip irrigation, before turning to the three case studies and, finally, a discussion based on the analytic framework, and conclusions.

Drip irrigation as a nexus idea for small farmers in drylands

Drip irrigation – also known as micro-, localized, or trickle irrigation – uses networks of pipes and tubes to apply water directly to the soil surface or root zone of plants. As a water-efficient technology, it has been promoted as a demand-side management option for reducing water consumption while maintaining yields, particularly through minimizing nonproductive evaporative losses (see e.g. Narayanamoorthy, 2004; Rijsberman, 2006). However, one review of the literature on drip irrigation has found that there are no common definitions of efficiency in use, and that measured efficiency gains can depend on specific boundary assumptions and operational conditions, including time and spatial scales (van der Kooij, Zwarteveen, Boesveld, & Kuper, 2013).

The pressurization of water used in drip irrigation has led to research on water–energy linkages. That the rapid expansion of groundwater irrigation has been underwritten by energy subsidies, which has in turn contributed to depletion of aquifers and placed strains on energy supply, is well documented (Scott & Shah, 2004; Shah, Scott, Kishore, &

Sharma, 2004). For example, research in Spain suggested that drip irrigation can be more energy-efficient than other technical options (Hardy, Garrido, & Juana, 2012); while research in China found that conversion from groundwater flood irrigation to drip irrigation can reduce both energy consumption and carbon emissions (Wang et al., 2012). An examination of trade-offs between water and energy in Australian irrigation concluded that drip irrigation is appropriate for reducing energy consumption in pumping groundwater, while surface water should be used as flood irrigation to avoid the energy costs of pressurization (Jackson, Khan, & Hafeez, 2010).

Drip irrigation therefore connects food production with water and energy use efficiency. Drip irrigation technology predates the WEF nexus concept, and the food–water, energy–water and energy–food dimensions have generally been considered separately rather than as a trilateral nexus. Nonetheless, the potential combined benefits of drip irrigation for these dimensions is well recognized, although questions remain on the necessary boundary conditions for them to be realized (see e.g. Jackson et al., 2010; van der Kooij et al., 2013). Experiences with adoption of drip irrigation in drylands can therefore serve as a model for understanding how nexus ideas might have positive or negative impacts on dryland farmers.

We propose a simple analytic framework with three components to assess whether drip irrigation has improved water, energy food security and benefited small farmers in Morocco. The first component asks whether small farmers have been able to adopt drip irrigation. The second component considers whether adoption of drip irrigation has improved water, energy or and food security, and how, and for who. The third component asks whether drip irrigation adoption has affected other markers of poverty, such as employment. In the discussion at the end of the article, this framework is used to consider what lessons can be drawn to help small farmers benefit from nexus ideas.

The development of drip irrigation in Morocco

Morocco is a water-scarce country, with unmet energy needs and an economy based on agriculture. Also, over recent decades it has experienced increasingly frequent and intense droughts. It therefore provides a useful case in which to consider interconnected issues of water, energy and food in a context of climatic variability and change.

Population growth and socio-economic development have increased demands for water and energy in Morocco. The national water potential, estimated at 22 km^3 per year, is above current demand estimates of 15.7 km^3 per year (Departement de l'Environnement, 2009; Droogers et al., 2012). However, demand and supply are not distributed evenly in time or in space. Precipitation is highly seasonal, with high intra- and inter-annual variability, and a large proportion of irrigation, industry and domestic demand is in areas with little rainfall, such as Casablanca. To better match supply and demand the government is developing water infrastructure for the mobilization, storage and transfer of water, but the investment costs are high because the majority of easy gains have already been made. For example, the 130 large dams constructed before 2009 had a combined storage of 17.5 km^3, yet the 60 dams planned by 2030, all combined, will add only 1.7 km^3 (Departement de l'Environnement, 2009). Population and economic growth will place further stresses on water resources, and expectations are that water deficits will increase markedly in the future, particularly if the impacts of climate change are taken into account (Droogers et al., 2012).

Energy supply and demand have risen steadily, but demand has grown faster. To meet energy demand, Morocco needs to invest approximately €1 billion per year for the next

10 years in new generation capacity (Vagliasindi, 2013). Growth of electricity demand has been particularly large, both from an increasing industrial base and due to a highly successful rural programme that increased electrification rates from 18% to 97% between 1995 and 2011. Around 96% of national energy needs are met through imports, which has created huge financial pressures on the state to reduce energy subsidies (Vagliasindi, 2013). Since 2009 the government has been phasing out fuel and energy subsidies, which has reduced public spending and increased consumer prices, although some support for diesel remains (Reuters, 2014).

Morocco is therefore a country in which both water and energy are scarce and require considerable investment to meet demand. Both water and energy are also significant resource inputs to Morocco's chief economic activity, agriculture.

Since the 1980s the mobilization and supply of both energy and water to agriculture has become a critical question. A series of intense droughts, unprecedented in the last 500 years, have struck the country since 1981, including the three successive dry years of 1999–2001 (Chbouki, Stockton, & Myers, 1995; Touchan et al., 2008). Relatively low use of irrigation meant that agriculture was highly sensitive to drought. Morocco also did not have extensive rural government services, subsidies, safety nets or markets; consequently, the social and economic impacts of the drought were severe, and helped precipitate a national economic crisis in the 1980s (Doukkali, 2005).

Officially, around 90% of Moroccan agriculture was rainfed, with the majority of irrigation schemes state-managed and based on surface water gravity and flood irrigation. However, in response to this onset of drought, and the increasing access to and affordability of rural electricity, diesel pumps and fuel, many farmers dug private tube wells and started pumping groundwater, often without permits. As a result, aquifers across Morocco have been rapidly depleted. For example, the Souss aquifer, a strategic reserve in southern Morocco, has been falling an average of 2 m per year for the last 30 years (Bouchaou et al., 2008). Groundwater pumping has also added strain to the energy sector and state energy subsidies, although that is rarely mentioned by comparison to the issue of water.

This is creating a dilemma for the state. On the one hand, there is a need to manage strategic water reserves and energy demand. On the other, agriculture is an engine of the economy and a social safety net, and it is difficult to enforce procedures and permits. The problem is complex, and linked to issues of social development, poverty alleviation, institutional reform and international trade, and not only to issues of water and energy scarcity and agricultural production and productivity.

Among other approaches, such as institutional reforms, drip irrigation has been promoted by the state as a technical solution. Increasingly large subsidies to support adoption of drip irrigation have been offered: there is currently a standard subsidy of 80%, rising to 100% for farms smaller than 5 ha. New forms of organization and coordination have also arisen for the management of irrigation and delivery of services. This has included community-based schemes, the empowerment of private companies to provide technical assistance alongside infrastructure sales, and also the world's first public–private partnership in irrigation: the Guerdane scheme in Taroudant Province in the Souss Massa (Errahj, Kuper, Faysse, & Djebbara, 2009; Houdret, 2012).

There has, then, been a large effort towards achieving the apparently obvious benefits of drip irrigation in terms of reducing water and energy consumption, boosting agricultural productivity and increasing drought tolerance. However, the case studies in the next section of this article suggest that it is not clear that drip irrigation is fulfilling its potential in Morocco as a technical adaptation to water stress and drought. This appears to be largely due to issues of institutions, policies and administration, resulting in barriers to the uptake of drip irrigation and unintended consequences of drip irrigation uptake.

The case studies

Bitit and Ain Chegag, Sebou

The 2200 km^2 Saiss sub-basin in the upper Sebou accounts for 11% of Morocco's annual water endowment. The mountainous Saiss sub-basin holds around 8000 commercial and subsistence farms, 37,000 ha of which are irrigated. Commercial farms oriented to international exports include the cultivation of water-intensive crops, with 4500 ha of apple orchards and hundreds of hectares of wine vineyards, particularly in the fertile valley floor (personal communication, director of Sefrou and el Hajeb Provincial Department of Agriculture, 23 February 2009). Of the irrigated land, 32% is flood-irrigated from surface waters, 45% is flood-irrigated from groundwater, and about 22% is drip-irrigated from different sources, mainly groundwater (Rhaouti, 2007).

Since the 1980s, the onset of more frequent drought and increasing access to international export markets has driven groundwater abstraction for agriculture. The number of wells has risen from less than a dozen to over 9000 in the last 40 years, while annual precipitation in the area has fallen by 33%, and the resulting imbalance has led to a water deficit of 108 m^3/y (Kalpakian et al., 2014). This has reduced flows in springs, and some sources and wells have dried up completely, particularly in the uplands. According to the Sebou River Basin Agency (ABHS), the Saiss aquifer fell 70 m over the 27 years before 2008, and on current trajectories could be exhausted within 20 years (Rhaouti, 2007). Against this background, the ABHS has prioritized the promotion of drip irrigation among policies to reduce water demand in the basin.

A knowledge, attitudes and practice survey of 519 farmers in the Ain Chegag and Bitit communities found that 83% of farmers identified drip irrigation as the best means of conserving water resources (Kalpakian et al., 2014). Farmers identified the key benefits as savings in water, energy, and labour inputs.

However, despite their recognizing the potential benefits of drip irrigation, only 10% of farmers had adopted it. The knowledge, attitudes and practice survey identified a complex set of institutional barriers to financing arising from issues of land tenure, the subsidy and credit system, and administrative requirements. For example, 48% of farmers identified a lack of money as the primary obstacle to adoption; 16% cited the administrative complexity of accessing the subsidy system; and 5% were impeded by land fragmentation or being tenant farmers. Other farmers preferred flood irrigation due to their proximity to reliable sources (Kalpakian et al., 2014).

At the root of several of these barriers is the diversity of land tenure systems in Morocco. Common forms of ownership include *melk* (private property), *'urfi* (informal, unregistered but communally recognized), *sulaliya* (tribal or communal title), *habous* (a religious endowment), *guich* (formerly used as compensation for military service) and *domain* (state land). In the past, possessors of communally owned lands, such as *'urfi* or *sulaliya*, were disqualified from obtaining mortgages because their properties were ultimately owned by the community. Even with large subsidies provided by the state, poor farmers would need access to credit to finance the remaining investment cost of drip irrigation. In Ain Chegag and Bitit, poor farmers were disproportionately holders of *'urfi* or *sulaliya* lands, and therefore without access to mortgages and credit. However, the laws have recently been changed, allowing mortgages for land for which legitimate use rights have been officially documented.

The knowledge, attitudes and practice survey also indicated that education was a major factor in being able to register land, with literate and illiterate farmers having very different levels of success in managing the required paperwork and accessing the bureaucracy. Similarly, although a 100% subsidy is available for small farmers (under 5 ha) to install drip irrigation, considerable paperwork and knowledge are required to access it. Private firms that sell drip irrigation systems generally offer farmers help with the paperwork and administration for credit and subsidies. However, they charge for the service, and many farmers distrust their motives.

Institutional fragmentation was also a challenge. The ABHS was charged with planning and allocating water at the basin level, and it tightly controlled permits to dig wells, out of concern with declining aquifer levels (also requiring evidence of land ownership). According to procedure, the Provincial Department of Agriculture required evidence of a well permit before providing farmers with subsidies for drip irrigation investments. However, the two agencies had only limited interaction and did not have harmonized practices, with the Provincial Department of Agriculture promoting irrigation and the ABHS attempting to restrict the number of wells. As well as education, physical distance from and between the location of the two offices was also a significant impediment to farmers' accessing these state services.

One clear implication of this case study is that the uptake of drip irrigation by small farmers was conditioned by administrative processes and regulations. Improved coordination in planning and operations between the involved agencies at a local level could conceivably have harmonized policies and fostered a strategic plan to promote drip irrigation within sustainable limits of groundwater abstraction.

However, it is also clear that there are important issues that go beyond policy and administrative coordination. In Ain Chegag and Bitit, the ability of farmers to access subsidies and take up drip irrigation depended on a wide range of factors, including the source of water being used, education level, form of property ownership, level of land fragmentation, and physical distance to administrative centres. Other social factors that researchers found to be significant barriers (all of which varied markedly between communities in a relatively small geographic area) included customary law, levels of social capital, and the specific obstacles faced by women farmers from some communities in inheritance law and accessing state services and subsidies. These barriers to small farmers' accessing the policy instruments (subsidies and credit) supporting drip irrigation are markers of social exclusion and poverty, and are much broader issues of development than a concern with resource use efficiency.

Lamzoudia, Tensift

The rural commune of Lamzoudia covers a plain of 77,000 ha south-west of Marrakech in Tensift Basin. It consists of 55 *douar* (small villages) totalling approximately 22,000 people. Traditional livelihoods in this area are based on sedentary agro-pastoralism, with pasture and rainfed winter barley being used to support sheep. The rangelands are predominantly state-owned land, although the area also includes 1725 ha of communally owned property managed through usufruct rights.

The climate is semi-arid, with irregular rainfall, concentrated in the winter months from October to March, and high evapotranspiration. Prior to the 1980s average annual rainfall was 250 mm/y, but since then a new precipitation regime has emerged that averages just 150 mm/y. Furthermore, whereas rain used to fall in regular, small quantities, it now falls in less frequent and more intense bursts that are less beneficial to barley production. These intense rains can cause flooding, and are thought to be less effective at recharging groundwater.

With greatly reduced productivity of barley and rangelands, the traditional rainfed agriculture-livestock system has been significantly negatively impacted. It has been partially replaced by agriculture irrigated with groundwater using the pumps and energy that became increasingly accessible during the 1980s.

This conversion to groundwater could have been used as an important buffer for farmers, granting time to adapt to new, drier conditions. However, in the absence of appropriate advice, and acting individually and tactically rather than strategically, the farmers of Lamzoudia made unsustainable choices. With improved access to groundwater and to markets for cash crops, many farmers converted to horticulture. Water-intensive crops, including watermelons and orchard fruits, are now key products of this area. Groundwater levels have fallen over 180 m in the last 40 years, and in 2010, the authors found that communities in the area were increasingly reporting dry or salinized wells.

Drip irrigation was particularly adopted by new commercial farmers, many of whom purchased land from destitute agro-pastoralists. For them, drip irrigation offered a means to reduce the energy costs of groundwater pumping. However, they also saw it as a means of managing costs, and of converting efficiency savings in water and energy into larger irrigated areas. The uptake of drip irrigation has therefore had no discernable impact on overall water conservation, although it has benefited users' agricultural productivity.

In contrast, smaller farmers tended to view the returns on drip irrigation as insufficient to justify costly investment. This was largely due to their difficulties in accessing subsidies due to their land status, as in the case study from Sebou. Their incentives for investment were further lowered due to the unbounded nature of groundwater: any water efficiency savings resulting from their investment in drip irrigation could be used by others without restriction.

The incentives on both commercial farmers and small traditional farmers have therefore been to convert water into cash as quickly as possible rather than using water efficiently, a classic "tragedy of the commons" scenario. Following the work of Elinor Ostrom (1990) and others, co-management or community-based management is a common prescription for common property resources. However, in Lamzoudia the community lacked the social capital and local institutions necessary to develop an agreement to regulate private abstractions from the common property aquifer within sustainable levels (which in any case would be hard to define). The presence of outside investors and commercial farmers made agreement even more unlikely.

The experience of Lamzoudia offers an example of Jevon's Paradox (Alcott, 2005), suggesting that, very simply, the adoption of technology offering water (and energy) use efficiency is not necessarily sufficient by itself to resolve scarcity. By comparison to flood irrigation from groundwater, drip irrigation offered greater water and energy efficiency. However, water users and farmers operated under incentives that did not encourage conservation, and they lacked access to the capital and institutions necessary to change those incentives.

Guerdane and Issen, Souss Massa

The Souss Massa basin covers 27,000 km^2 in the south of Morocco. Producing more than 50% of Morocco's vegetable and citrus exports, it is one of the most important agricultural regions in the country. This export-based agriculture has grown rapidly since economic liberalization in the 1980s.

Since the 1980s, average annual precipitation has fallen to 80% of the long-term mean, and variability has also markedly increased. Traditionally around 80% of the area was rainfed, and focused on olive and cereal production. However, increasing water variability and scarcity, and demand for water intensive crops for export, have driven demand for irrigation water (Keith & Ouattar, 2004; Van Cauwenbergh & Idllalene, 2012). Around 93% of water in the Souss Massa is now used in agriculture.

Surface water has been mobilized for irrigation through the construction of eight medium-to-large dams. However, agricultural water demand has continued to grow in areas unsupplied by these irrigation schemes, with farmers turning to pumped groundwater. Within areas supplied by irrigation schemes, groundwater is also used for supplementary irrigation. In total, almost 70% of the water used in irrigation is drawn from groundwater, and the Souss aquifer has a deficit of 360×10^6 m^3/y. The aquifer has fallen at a rate of 1.5 m to 2 m per year, and in some places tubewells are now at over 200 m (Houdret, 2012).

As elsewhere in Morocco, the state has responded with a variety of measures that have included the promotion of drip irrigation. This has included the provision of pressurized surface water, as well as measures to promote drip irrigation to reduce groundwater abstraction. One of these schemes is the Guerdane irrigation project, a public–private partnership (PPP) managed by the Amensous company. This scheme delivers water to an area which is prioritized for the production of citrus for export. Previously, the farms in this area were using $10,000$ m^3 ha^{-1} y^{-1} for citrus production, all abstracted from groundwater for flood irrigation. The new irrigation scheme requires drip irrigation, for which 6000 m^3 ha^{-1} y^{-1} is sufficient for the same yield of citrus. Of this 6000 m^3, two-thirds is provided from surface water delivered by Amensous, while one-third is expected to be pumped from groundwater. This represents a potential saving of ~ 8000 m^3 ha^{-1} y^{-1} in terms of groundwater abstraction (personal communication, Amensous officials, 21 May 2014). Although the energy cost of pressurized delivery of surface water to the scheme was not available, it can be assumed to be substantially less than the cost of pumping from the deep aquifer, so this water saving implies concomitant energy savings.

Unusually for Morocco, where approximately 70% of farms are smaller than 2 ha, the Guerdane scheme serves an area where 67% of farms are larger than 20 ha. The owners of these larger farms are generally wealthy, and for many either agriculture is a secondary income or the land is rented to a foreign company. In addition to other benefits for property owners, the values of land in the PPP perimeter have risen from MAD 25,000 (USD 2800) to MAD 250,000 (USD 28,000) per hectare since the scheme's inception in 2009 (personal communication, Amensouss and Ministry of Agriculture officials, 21 and 22 May 2014). The Guerdane PPP scheme has therefore been criticized for subsidizing the infrastructure, productivity and capital gains of larger, wealthy farmers at the taxpayer's expense (Houdret, 2012; Van Cauwenbergh & Idllalene, 2012). Other criticisms have focused on the impact on the traditional water rights of the relatively poor upstream communities affected by the mobilization of surface water for the Guerdane scheme (Houdret, 2012). There are also questions about the cultivation of citrus for export, as orchard crops fix water demand for years or even decades rather than being able to respond flexibly to annual variation in supply.

At the nearby Issen traditional scheme, several farmers reported that they view reduced labour costs, rather than increased water and energy efficiency, as the main benefit of converting from flood to drip irrigation. Rather than needing to employ part-time labourers to channel irrigation waters, the drip irrigation system allowed small farmers to irrigate alone and with minimal effort. Other farmers increased their irrigated area, resulting in smaller-than-expected net water savings (or none). Some farmers have both increased the irrigated area and reduced their labour force. Others have converted to water-intensive crops such as fruit and vegetables, with further consequences for water demand.

Discussion

Have small farmers been able to adopt drip irrigation?

In each case study, wealthier farmers were better able than small farmers to adopt drip irrigation. Although the state offers subsidies of up to 100% to offset the high initial investment costs, many small farmers experienced significant barriers to accessing this support.

One of the most significant issues was land tenure status. Small farmers on private property (*melk*) in Souss Massa were better able to access subsidies than farmers on other forms of land tenure, such as '*urfi* or *sulaliya*, in Tensift and Sebou. Other barriers to accessing subsidies and services involved literacy and ability to work through administrative processes, geographic remoteness from administrative offices, gender, and ability to procure a well permit from water management authorities. While the subsidy of 80% or 100% was an effective means of supporting uptake of drip irrigation by small farmers on private property, it did not appreciably benefit small farmers with non-private land tenure or from the most marginalized groups.

Other small farmers did not see sufficient economic reasons to invest in drip irrigation. Upstream farmers in Sebou and the Issen perimeter in Souss Massa, with more assured surface water supply, had few incentives to conserve water to benefit downstream users. Similarly, some smallholders in Tensift saw little value in water conservation efforts that would be open to abuse from free riders in a common pool resource. Another economic barrier for some farmers in Sebou and Tensift were farms that were too small, or too fragmented, to realize benefits from drip irrigation, particularly where they were not able to develop cooperative agreements with their neighbours.

These findings suggest that policies supporting the adoption of technical options may not be successful without careful anticipation of institutional barriers. Many of these issues – land tenure, education and gender – are fundamental to dimensions of poverty, are complex and diverse, and are frequently highly political and difficult to resolve. Technical options inspired by nexus ideas, and the policies supporting them, are highly unlikely to initiate transformative changes that resolve these issues. However, policy formulation processes could potentially map what barriers there might be to adoption by small farmers, and how they might be lessened, and ensure that policies are framed appropriately.

Has drip irrigation improved water, energy or food security?

The three cases demonstrate that the efficiency savings offered by drip irrigation do not necessarily reduce on-farm consumption of water or energy, and have not had a clear positive impact on water, energy or food security. Rather, the positive and negative impacts of drip irrigation on water, energy and food security appear to be distributed among different groups and at different scales.

Whether drip irrigation generates significant water savings in any real sense at the hydrological scale has been challenged by many authors (see e.g. Perry, 2007; van der Kooij et al., 2013; Ward & Pulido-Velazquez, 2008). Efficiency savings can be generated by reducing non-beneficial evapotranspiration, losses to sinks such as the sea or the desert, and improved irrigation management. However, aside from these potential savings, excess irrigation water not used in beneficial evapotranspiration (contributing to plant growth) returns to the aquifer, where it becomes available for use by others. In principle, replacing flood with drip irrigation could maintain or improve aquifer levels if water demand were kept constant. However, if farmers use efficiency savings to increase water demand, by expanding the irrigated area or adopting water-intensive crops, there may be no net benefit to aquifer levels. In principle, wide adoption of drip irrigation without control of water demand could reduce the amount of water available in the aquifer for other users (Ward & Pulido-Velazquez, 2008).

Large reductions in excess irrigation water by adoption of drip systems, therefore, do not necessarily translate into real savings. To restore aquifer levels, water efficiency savings need to be translated into permanent reductions in abstraction. However, there is always a pressure to use water for economic activities rather than for environmental flows or for non-economic uses such as subsistence agriculture by the poor. Indeed, the Souss Massa and Tensift cases found that the incentives operating on farmers encouraged many of them to expand irrigated areas and cultivate water-intensive crops rather than reduce groundwater abstractions. In the Sebou case, the majority of farmers were aware of the potential benefits to water and energy conservation. However, it appeared that at least some of the minority who had adopted drip irrigation were, like those in Tensift and Souss Massa, adopting water-intensive crops.

Energy efficiency can also be improved by drip irrigation (see e.g. Hardy et al., 2012; Wang et al., 2012). That might be of more significant value to the state where energy is heavily subsidized, or to the farmer where energy costs are high. However, energy consumption does not decline unless the total quantity of water pumped is also reduced. If water efficiency gains from drip irrigation are used to extend irrigation areas and increase water intensity rather than reduce water consumption, total energy consumption will not decline either. Replacing flood irrigation from surface water with drip irrigation does introduce energy costs for pressurization to the farmer, but here too these can also be offset – and profit margins increased – by growing water-intensive, commercial crops.

In these case studies, the farmers adopting drip irrigation have enlarged their incomes by maximizing the profitability of their water use. From their perspective, water and energy costs are less significant than profit margins. Instead, gains have been made in agricultural productivity and income generation, and benefits have tended to accrue directly to the farmer rather than to society in general.

Has drip irrigation impacted on other aspects of poverty?

The case studies indicate that small farmers who adopted drip irrigation increased agricultural productivity, often through switching to new crops. For these farmers, adopting drip irrigation appeared to have positive impacts on income and other markers of poverty.

Drip irrigation requires less labour inputs than flood irrigation. Several small farmers reported that this had increased their personal quality of life by reducing work hours. Both large farmers and small farmers had been able to reduce their costs for agricultural labour. While this increased the profit margins of farmers, it necessarily means a reduction in employment for agricultural labourers. This implies that adoption of drip irrigation can contribute to concentrating the economic benefits of agriculture in the hands of property owners, while increasing poverty among agricultural labourers, thus making the rural economy less inclusive. This finding may be context-dependent, however; there was some evidence that expansion of drip irrigation on new commercial farms in formerly rainfed areas of Tensift has provided employment opportunities, particularly for women.

Aside from employment issues, the use of surface water resources for the Guerdane scheme has reportedly also resulted in distributional impacts, with upstream communities in the upstream area losing access to water sources (Houdret, 2012).

There are, therefore, questions to be asked about who has benefited and who has lost from the adoption of drip irrigation and the associated public subsidy. There is some evidence that, at least in some cases, drip irrigation technology and subsidies have exacerbated social and economic inequality. Individual farmers, particularly those who were already wealthy, have been the largest beneficiaries, while the impact on agricultural labourers has been negative in some instances. In particular, questions can be asked about the 80% public subsidy available to large commercial farmers for adopting drip irrigation. They are more likely to have the capital necessary for investment, and are more likely to respond to market signals and reduce energy and water costs to maximize profits.

This does not mean that drip irrigation and the associated public subsidies are entirely negative. However, those developing policies to support technical options need to be aware that creating winners and losers is more likely than creating a "win-win" scenario, and that technical options can exacerbate social and economic inequalities.

Can nexus ideas be made to work for poor drylands farmers?

Reflecting on these cases of drip irrigation adoption, on how it has affected water, energy and food security, and on some of the impacts it has had on poverty provides two important lessons for understanding how nexus ideas might benefit small dryland farmers.

The first lesson is that, while drip irrigation has clearly benefited some, the realization of benefits depends on a number of contextual factors. Farmers' access to drip irrigation depends on their available capital, perceived returns on investment, and access to subsidies. In turn, these depend on markers of poverty and social marginalization such as farm size and land tenure. Similarly, drip irrigation does not necessarily result in greater water or energy security at the basin level due to the incentives acting on individual farmers. This suggests that the conditions for successful application of nexus ideas require careful assessment, and that supporting policies should be targeted at specific user groups in specific contexts.

The second lesson is that drip irrigation has the potential to create winners and losers between societal groups and at different scales, and that this is more likely than creating positive outcomes for all. This lesson implies that, if policy based on the WEF nexus is to benefit small farmers, it should purposefully integrate *ex ante* evaluation of distributional impacts and engage with pro-poor development agendas.

More fundamentally, we would ask to what end WEF nexus approaches are oriented. Is the reduction of trade-offs between water, energy and food security considered an end in itself, or does it support higher-level social goals such as the reduction of poverty? The most significant water, energy and food security challenges faced by poor dryland farmers are to do with availability, access and stability of those resources, not efficiencies of use or optimization of trade-offs between them. In large areas of rural Morocco, food production and income generation depend on scarce and variable water supplies. Pressurizing surface water for drip irrigation in Sebou and Issen might not be an efficient use of energy resources (Jackson et al., 2010), but does deliver crucial water supplies that help alleviate rural poverty and strengthen resilience to drought. By contrast, pressurization of surface waters for drip irrigation in Guerdane might be less energy-intensive than pumping groundwater resources, but has reportedly impacted the water rights of marginalized upstream communities. The fundamental issues in these cases are to do with poverty and marginalization, not efficient energy use.

Addressing basic insecurity in water, energy and food, and other dimensions of poverty such as rights, income, employment, health and education, are the focus of international development efforts. Where opportunities can be identified to reduce trade-offs between these types of security, that is all to the good. However, we are not convinced that this is a beneficial starting point for analysis if it might lead to situations in which poor farmers are expected to prioritize efficiency over fundamental security and poverty reduction.

Conclusions

This article has presented findings from three sets of case studies on the uptake of drip irrigation in Morocco. Although drip irrigation predates the current formulation of the WEF nexus concept, it has long been recognized that it offers water and energy efficiency savings while maintaining, or even increasing, agricultural production and productivity.

The case studies suggest that making nexus solutions and technologies work for small farmers in drylands is complicated. Institutional barriers to access can be complex and diverse, and incentives may not always be sufficient to encourage users to save water or energy. The boundary conditions for achieving desired outcomes might be highly specific. On the other hand, the prioritization of technical and policy options for resource use efficiency has, in some cases, the potential for unintended consequences that include the creation of winners and losers at different scales and between different groups.

Based on these case studies, we suggest that those interested in developing WEF nexus ideas to benefit small farmers in drylands ensure that boundary conditions for success are well understood, that supporting policies are carefully targeted, that *ex ante* evaluations consider potential distributional impacts, and that the technical, institutional and policy options proposed are supportive of pro-poor agendas of inclusive social and economic development.

Acknowledgements

We thank two anonymous reviewers for their efforts in critically reading the manuscript and making beneficial suggestions.

Funding

This work was supported by Canada's International Development Research Centre and the United Kingdom's Department for International Development through the Climate Change Adaptation in Africa programme: Mechanisms of Adaptation to Climate Change in the Rural Communities of Two Contrasting Ecosystems in the Plains and Mountains of Morocco [Project 104153] and Using Demand Side Management to Adapt to Water Scarcity and Climate Change in the Saiss Basin [Project 105439]. This article also draws on fieldwork carried out in May 2014 and supported by the Future Agricultures Consortium.

Disclosure statement

No potential conflict of interest was reported by the authors.

References

Alcott, B. (2005). Jevons' paradox. *Ecological Economics*, *54*, 9–21. doi:10.1016/j.ecolecon.2005.03.020

Allan, J. A. (1997). "Virtual water": A long-term solution for water-short Middle Eastern economies? The 1997 British Association Festival of Science, Roger Stevens Lecture Theatre, University of Leeds, Water and Development Session.

Allan, J. A. (2003). Virtual water - the water, food, and trade nexus. Useful concept or misleading metaphor?. *Water International, 28*, 106–113. doi:10.1080/02508060.2003.9724812

Allan, J. A., Keulertz, M., & Sojamo, S. (Eds.). (2012). *Handbook of land and water grabs in Africa: Foreign direct investment and food and water security*. Abingdon: Routledge.

Batliwala, S. (1982). Rural energy scarcity and nutrition: A new perspective. *Economic and Political Weekly, 17*(9), 329–333.

Bizikova, L., Roy, D., Swanson, D., Venema, H. D., & McCandless, M. (2013). *The water-energy-food security nexus: Towards a practical planning and decision-support framework for landscape investment and risk management*. Winnipeg: International Institute for Sustainable Development.

Bouchaou, L., Michelot, J. L., Vengosh, A., Hsissou, Y., Qurtobi, M., Gaye, C. B., & … Zuppi, G. M. (2008). Application of multiple isotopic and geochemical tracers for investigation of recharge, salinization, and residence time of water in the Souss-Massa aquifer, southwest of Morocco. *Journal of Hydrology, 352*, 267–287. doi:10.1016/j.jhydrol.2008.01.022

Chbouki, N., Stockton, C. W., & Myers, D. E. (1995). Spatio-temporal patterns of drought in Morocco. *International Journal of Climatology, 15*, 187–205. doi:10.1002/joc.3370150205

Departement de l'Environnement. (2009). *Strategie nationale de l'eau*. Rabat: Ministry of Energy, Mines, Water and Environment.

Doukkali, M. R. (2005). Water institutional reforms in Morocco. *Water Policy, 7*, 71–88.

Droogers, P., Immerzeel, W. W., Terink, W., Hoogeveen, J., Bierkens, M. F. P., Van Beek, L. P. H., & Debele, B. (2012). Water resources trends in Middle East and North Africa towards 2050. *Hydrology and Earth System Sciences, 16*, 3101–3114. doi:10.5194/hess-16-3101-2012

Errahj, M., Kuper, M., Faysse, N., & Djebbara, M. (2009). Finding a way to legality, local coordination modes and public policies in large-scale irrigation schemes in Algeria and Morocco. *Irrigation and Drainage, 58*, S358–S369. doi:10.1002/ird.526

Greeley, M. (1987). Energy and agriculture: interactions and impact on poverty. *IDS Bulletin, 18*, 47–54. doi:10.1111/j.1759-5436.1987.mp18001007.x

Hardy, L., Garrido, A., & Juana, L. (2012). Evaluation of Spain's water-energy nexus. *International Journal of Water Resources Development, 28*, 151–170. Retrieved from Google Scholar10.1080/07900627.2012.642240

Hoff, H. (2011). *Understanding the nexus. Background paper for the Bonn2011 conference: The water, energy and food security nexus*. Stockholm: Stockholm Environment Institute.

Houdret, A. (2012). The water connection: Irrigation, water grabbing and politics in southern Morocco. *Water Alternatives, 5*, 284–303.

Jackson, T. M., Khan, S., & Hafeez, M. (2010). A comparative analysis of water application and energy consumption at the irrigated field level. *Agricultural Water Management, 97*, 1477–1485. doi:10.1016/j.agwat.2010.04.013

Kalpakian, J., Legrouri, A., Ejekki, F., Doudou, K., Berrada, F., Ouardaoui, A., & Kettani, D. (2014). Obstacles facing the diffusion of drip irrigation technology in the Middle Atlas region of Morocco. *International Journal of Environmental Studies, 71*, 63–75. doi:10.1080/00207233.2014.881956

Keeney, R. L., & Wood, E. F. (1977). An illustrative example of the use of multiattribute utility theory for water resource planning. *Water Resources Research, 13*, 705–712. doi:10.1029/WR013i004p00705

Keith, J. E., & Ouattar, S. (2004). Strategic planning, impact assessment, and technical aid: the Souss-Massa Integrated Water Management Project. *Journal of Environmental Assessment Policy and Management, 6*, 245–259. doi:10.1142/S1464333204001699

Narayanamoorthy, A. (2004). Drip irrigation in India: Can it solve water scarcity? *Water Policy, 6*, 117–130.

Ostrom, E. (1990). *Governing the commons: The evolution of institutions for collective action*. Cambridge: Cambridge University Press.

Perry, C. (2007). Efficient irrigation; inefficient communication; flawed recommendations. *Irrigation and Drainage, 56*, 367–378. doi:10.1002/ird.323

Rasul, G. (2014). Food, water, and energy security in South Asia: a nexus perspective from the Hindu Kush Himalayan region. *Environmental Science & Policy*, *39*, 35–48. doi:10.1016/j.envsci.2014.01.010

Reuters. (2014, January). Morocco ends gasoline, fuel oil Subsidies. *Reuters News Agency*.

Rhaouti, S. (2007). *Presentation at the International Workshop on Adapting to Water Scarcity and Climate Change in North Africa: DSM and Its Implications for Social Equity and Environmental Sustainability*. Al Akhawayn University, Ifrane, Morocco, 3–4 December 2007.

Rijsberman, F. R. (2006). Water scarcity: Fact or fiction? *Agricultural Water Management*, *80*, 5–22. doi:10.1016/j.agwat.2005.07.001

Sachs, I. (1984). *Energy and agriculture: Interaction futures policy implications of global models* (pp. 25–40). New York, NY: Harwood Academic.

Scott, C. A., & Shah, T. (2004). Groundwater overdraft reduction through agricultural energy policy: Insights from India and Mexico. *International Journal of Water Resources Development*, *20*, 149–164. doi:10.1080/0790062042000206156

Shah, T., Scott, C., Kishore, A., & Sharma, A. (2004). *Energy-irrigation nexus in South Asia: Improving groundwater conservation and power sector viability* IMWI Research Report no. 70. Colombo: International Water Management Institute.

Siddiqi, A., & Anadon, L. D. (2011). The water-energy nexus in Middle East and North Africa. *Energy Policy*, *39*, 4529–4540. doi:10.1016/j.enpol.2011.04.023

Touchan, R., Anchukaitis, K. J., Meko, D. M., Attalah, S., Baisan, C., & Aloui, A. (2008). Long term context for recent drought in Northwestern Africa. *Geophysical Research Letters*, *35*, L13705, doi:10.1029/2008GL034264

Trieb, F., & Müller-Steinhagen, H. (2008). Concentrating solar power for seawater desalination in the Middle East and North Africa. *Desalination*, *220*, 165–183. doi:10.1016/j.desal.2007.01.030

Vagliasindi, M (2013). *Implementing energy subsidy reforms: Evidence from developing countries*. Washington, DC: World Bank.

Van Cauwenbergh, N., & Idllalene, S. (2012). Opportunities and challenges for investment in Morocco. In J. A. Allan, M. Keulertz, S. Sojamo, & J. Warner (Eds.), *Handbook of land and water grabs in Africa: Foreign direct investment and food and water security* (pp. 193–206). Abingdon: Routledge.

van der Kooij, S., Zwarteveen, M., Boesveld, H., & Kuper, M. (2013). The efficiency of drip irrigation unpacked. *Agricultural Water Management*, *123*, 103–110. doi:10.1016/j.agwat.2013.03.014

Wang, J., Rothausen, S. G. S. A., Conway, D., Zhang, L., Xiong, W., Holman, I. P., & Li, Y. (2012). China's water-energy nexus: greenhouse-gas emissions from groundwater use for agriculture. *Environmental Research Letters*, *7*, 014035, doi:10.1088/1748-9326/7/1/014035

Ward, F. A., & Pulido-Velazquez, M. (2008). Water conservation in irrigation can increase water use. *Proceedings of the National Academy of Sciences*, *105*, 18215–18220. doi:10.1073/pnas.0805554105

Waughray, D. (2011). *Water security: The water-food-energy-climate nexus*. Washington, DC: Island Press.

Zhu, T., Ringler, C., & Cai, X. (2007). Energy price and groundwater extraction for agriculture: exploring the energy-water-food nexus at the global and basin levels. *International Conference on Linkages Between Energy and Water Management for Agriculture in Developing Countries, Hyderabad, India, 29–30 January 2007*.

Virtual-water content of agricultural production and food trade balance of Tunisia

Jamel Chahed[a], Mustapha Besbes[a] and Abdelkader Hamdane[b]

[a]Université Tunis El Manar, Ecole Nationale d'Ingénieurs de Tunis, Tunisia; [b]Université de Carthage, Institut National Agronomique de Tunis, Tunisia

This article is devoted to the assessment of Tunisian agricultural production and food trade balance water-equivalent. A linear regression model relating annual rainfall to crop yields is developed to estimate the agricultural production water-equivalent. Its implementation is based on national data for crop and animal production, leading to food demand water-equivalent quantification. Results highlight the relationship between agricultural and water policies and provide a picture of food security in the country in relation to local agricultural production, and to virtual water fluxes related to foodstuffs trade balance.

Introduction

Mobilization, transfer and water management programmes during the last five decades have greatly increased the availability of Tunisia's water resources. The essential goal of this hydraulic policy was simple: It aims at the total development of all surface water and groundwater resources in order to sustain the socio-economic development of the country and to promote a modern agriculture so as to increase local foodstuffs production and promote exports (Besbes, Chahed, & Hamdane, 2014). But the expected continuously growing water demand raises the question of national water security in the future.

Water security can be simply defined as sustainable access to water in sufficient quantity and acceptable quality. The water security issue covers primarily the security of drinking water supply; but it is also a food safety issue which is inseparable from agricultural policy and from national food balance. Irrigation is the most important water consumer, accounting for about 70% of the world's freshwater withdrawals (FAO, 2014). In arid countries, the two components of agriculture (rainfed and irrigated) play complementary roles in food security. Irrigation contributes to increasing and stabilizing local agricultural production and plays a key role in the promotion of rural areas. Rainfed agriculture plays a key role in food security and in food trade balance. The relationship between water security and food security is particularly important in the African context, where the economies rely heavily on agriculture (Besada & Werner, 2015). This dialectical relationship necessarily involves a holistic approach to water problems that takes into account multiple factors.

In conventional water resources planning, neither the water contribution to rainfed agriculture nor the virtual water associated with foodstuffs trade is taken directly into account. In fact, water policies rely on a classical hydraulic approach which considers only the potential of exploitable blue water. Such a focus excludes or underestimates other important resources. A more comprehensive approach not bound by the hydraulic approach is needed. According to Falkenmark (2001), consideration should be given to all kinds of water: both blue water supporting humanity and aquatic ecosystems, and green water supporting terrestrial ecosystems, agriculture and forestry. The potential of green water resources components is often poorly known. Given the importance of these resources, development of specific methods to identify and estimate their potential is important. On the other hand, at a local scale, it is important to collect field data to promote optimal use of the local water resources. For example, the work of Redwood, Bouraoui, and Houmane (2014) related to food production in peri-urban areas showed that investments in rainwater harvesting and greywater treatment at the farm level can increase the financial feasibility of peri-urban farms.

Many studies have been devoted to the evaluation of the amounts of water required in foodstuffs production (Hoekstra & Chapagain, 2006; Hoekstra & Mekonnen, 2011; Renault & Wallender, 2000; Zimmer, 2013). The virtual water content or water-equivalent of a product is measured (in cubic metres of water per tonne of food) at the place where it is produced. This article attempts to evaluate the amount of the water-equivalent of foodstuffs production; for this purpose a global linear regression model is developed. After validation, the model is used to determine the water-equivalent of cereals and olive oil production and the potential of rainfed agriculture (green water) and irrigated crops (blue water). An assessment of the water-equivalent of all agricultural production (crop and animal) in Tunisia is then performed. After estimating the trade balance of foodstuffs water-equivalent (virtual water), an assessment of food demand water-equivalent is produced and its variation analyzed. All these estimates highlight the relationships between water security and food security issues, in a holistic vision including all kinds of water resources: blue water, green water and virtual water.

Water resources, agriculture and green water

Tunisia is split between two climate regions: a Saharan climate in the south and a Mediterranean one in the northern region. The average annual rainfall is around 220 mm, corresponding to 36 km³/y of precipitation throughout the country. This average masks huge disparities: (1) spatial heterogeneity, with 1000 mm/y in the north and less than 100 mm/y in the south (Figure 1); and (2) temporal variability—minimum rainfall is one-third the average, and maximum is up to three times the average (Besbes, Chahed, Hamdane, & De Marsily, 2010).

Direct water needs (for drinking and for industrial and touristic activities) are estimated at 50 m³/y per capita, whilst the amount needed for food production is more than 1500 m³/y per capita (Besbes et al., 2010). Despite its limited resources, Tunisia has shifted towards mobilizing most of its water resources by way of a large water infrastructure. The total mean blue water potential is estimated at 4.8 km³/y, of which 2.7 km³/y are related to surface runoff and 2.1 km³/y represent groundwater resources. Irrigation is by far the largest water user and accounts for around 80% of the water demand.

During the last four decades, Tunisia has implemented a pragmatic policy to increase local food production and promote exports. Irrigation has been a crucial factor in the increase of agricultural production: The total area equipped for irrigation was multiplied by seven over the past four decades and is developing over 410,000 ha—still less than 10% of the total utilized agricultural area (Besbes et al., 2014). Irrigation is largely practised for

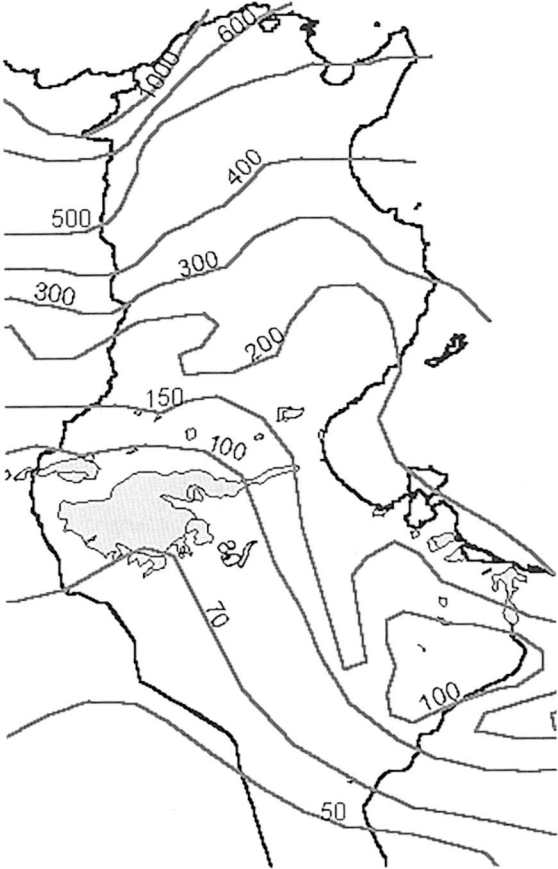

Figure 1.   Average isohyets in Tunisia (mm/y).

vegetable crops in various parts of the country, fruit trees (particularly citrus fruits) in the region of Cap Bon, and date palms in the oasis of the south, and secondarily for cereal crops and fodder in the north and the centre of the country. The irrigation sector contributes to a third of the agricultural production value (Hamdane, 2007); the other two-thirds come from rainfed agriculture, the production of which is directly related to rainfall, unevenly distributed in space and subject to strong seasonal and inter-annual variations.

The notion of 'green water' refers to rainwater consumed (Hoekstra & Chapagain, 2006). This corresponds, for a given agricultural production, to the amount of rainwater consumed during the growing period of the crop. The limits of the classical hydraulic water vision in addressing food security challenges in arid and semi-arid countries have been highlighted in previous articles (Besbes et al., 2010; Chahed, Besbes, & Hamdane, 2011). Furthermore, the concept of 'water security' versus 'water stress' derives from the same hydraulic approach and considers only the potential of water withdrawal. This approach is not complete, in particular where, as in Tunisia, the water resources involved in rainfed agriculture production are much more important than water used in irrigated agriculture (Chahed, Hamdane, & Besbes, 2008).

Mekonnen and Hoekstra (2011) present a global analysis of the water footprint related to agricultural production at the worldwide scale. They find that for most of the crops, the contribution of green water footprint toward the total consumptive water footprint (green and blue) is more than 80%. It should be remembered that these estimates are based on global modelling; it would be very welcome if case studies could specify these results at national, regional, and local levels. On the other hand, green water plays an important part in virtual water trade (cereals, oil, etc.), and its contribution is often decisive. Aldaya, Allan, and Hoekstra (2010) stressed in this respect the strategic importance of international green water trade and its contribution to water security, especially in water-scarce countries.

The Tunisian strategy of agricultural development has given priority to the irrigation sector. Rainfed agriculture has not received the same attention and has not had comparable progress. The area cultivated under rainfed conditions (about 4 million ha) is variable because of random rainfall conditions (Besbes et al., 2014). The two major farming systems, rainfed cereal crops and orchards (mainly olive trees), occupy more than two-thirds of arable land in different soil environments, more or less favourable for intensification. The grain agriculture lies in the north, while olive trees are grown in the central and southern regions. Whilst rainfed agriculture has not registered significant expansion in recent decades, the sector has improved its performance: Average yield for cereals has more than doubled over the past 30 years, and there is still significant margin for further improvement (Rezgui, Ben Mechlia, Bizid, Kalboussi, & Hayouni, 2005).

Nevertheless, the virtual water content of foodstuffs produced by rainfed agriculture is not well known. We have first to elaborate accurate methods to estimate water and soil resources—their potential, spatial distribution and variability, and their importance in the global foodstuffs balance. This also applies to virtual water flows associated with international foodstuffs trade, since a significant proportion of foodstuffs exchange comes from rainfed agriculture.

Water-equivalent of agricultural production and food demand

Modelling the water-equivalent of agricultural production

Several recent studies have studied variability in the production of rainfed agriculture by focusing their analysis on the relationship between rainfall and crop yields and on the vulnerability of plantations to successive episodes of drought (Gargouri et al., 2008; Ghrab, Gargouri, & Ben Mimoun, 2008; Rezgui et al., 2005). Tests conducted on wheat crops for seven agricultural seasons (1992/1993–1998/1999) in five different bioclimatic sites in Tunisia under rainfed and irrigated conditions indicate that grain yields are linearly correlated with net water consumption measured on the field (Rezgui et al., 2005). On the other hand, these experimental results suggest a minimum water consumption, at which grain production becomes possible, of around 210 mm/y. Presumably this threshold, which includes rainfall during the dry season, may vary from region to region. The threshold also includes the part of rainfall that feeds blue water systems, including the surface water and groundwater systems. For simplicity, one can admit that for a given region, the threshold is independent of the year. This hypothesis is quite plausible because of the relatively limited size of the regions. The experimental results of Rezgui et al. (2005) suggest that it is possible to calculate the volume of green water by performing a linear regression of grain production by annual rainfall accumulation during the production cycle in a given region. When the hydrological year, which begins in September, recovers all the production cycle, one may consider that rainfall accumulation will correspond to the annual rainfall during the hydrological year, as for cereal crops. This is not the case for olive trees, whose production cycle covers two successive hydrological years: part of the production year, and part of the preceding year. In order to consider the biannual alternate-bearing cycle of olive trees, one may consider that the two successive years are involved in the production of year n with two respective weights, α and β, such that $\alpha + \beta = 1$.

Based on empirical results obtained for cereal crops, an assessment model has been developed for systematic evaluation of the water-equivalent of rainfed agriculture production (Chahed et al., 2011). The basic assumption of the model is the linear relationship between rainfed agriculture production and rainfall during the production cycle. Thus, the model assesses the increase of agricultural production per unit of additional water volume. The amount of green water involved in the production of rainfed agriculture during hydrological year n is expressed in productive rainfall (volume per unit area) in a given region. This productive rainfall is expressed as the difference between the total rainfall and the threshold level of rainfall below which there is no production:

$$h_v(n) = (h_r(n) - h_s) \tag{1}$$

where $h_r(n)$ is the rainfall of hydrological year (n) and h_s is the rainfall limit below which the yield of year n, $R_v(n)$, is zero. It follows that $R_v(n)$ is directly proportional to h_v. Choosing to express R_v in quintals per hectare and h_v in mm, one can write:

$$h_v(n) = h_r(n) - h_s = 10 \times EE_p \times R_v(n) \tag{2}$$

where EE_p is the specific green water-equivalent in m^3 per kg of agricultural product in a specific region. The same analysis extended to all regions of the country leads to similar results. The green water volume nationwide is naturally expressed as the sum of regional contributions. This is equivalent to applying a spatial averaging operator ($-$) by weighting the quantities by the part of cultivated area in each region. Moreover, if one assumes that the part of cultivated area in each region is equal to its inter-annual mean value, one can determine the average amount of water green volume at the national scale. The corresponding spatio-temporal operator will be denoted as $< >$. The green water-equivalent volume (in cubic meters) of rainfed agriculture, which occupies an average area $<S>$ (in hectars) can be obtained using the two following formulations given by Equation (3) and Equation (4):

$$< \overline{EV_1} > = 10(< \overline{h_r} > - \overline{h_s}) < S > \tag{3}$$

$$< \overline{EV_2} > = 100\overline{EE_p} < \overline{R_v} >< S > \tag{4}$$

The notion of water-equivalent of agricultural production as defined above is similar to the notion of yield response to water (YRW), introduced by Doorenbos and Kassam (1979). Despite the simplicity of its formulation, YRW captures the essence of the complex linkages between production and water use by a crop. The UN Food and Agriculture Organization has largely contributed to the widespread dissemination of this concept. Today, YRW is widely used worldwide for a broad range of applications (Steduto, Hsiao, Fereres, & Raes, 2012).

As for rainfed crops, the water equivalent of irrigated crop production (in m^3) is proportional to the cultivated area $S_I(n)$, so that it is possible to express the amount of water (per unit area) used by irrigated crops as the sum of (1) $h_I(n)$, the net water used by irrigation (blue water), and (2) $h_v(n)$, the green water evaluated as for rainfed agriculture.

Letting $R_I(n)$ be the yield of irrigated crops, one may write:

$$h_1(n) + h_v(n) = 10 \times EE_p \times R_1(n) \tag{5}$$

This model, which uses annual regionalized data related to rainfall and production yields, has been extended to the national scale, giving relationships similar to those observed in each region. These relationships link rainfall and production yields; both are weighted by the regional harvested area. The agricultural statistics data (MARH, 1998–2007) allow the calculation of rainfall height and weighted values of crop yields. Thus, we may check whether these relationships have statistical meaning for cereal and olive tree farming in Tunisia. The results may then be used in a systematic assessment of the water resources involved in the production of these two types of farming products, and the method can be extended to the evaluation of the mean amount of green water resources. The results can provide a useful indication of the order of magnitude of green water resources involved in food production.

The linear regression model was first compared to a hydrologic soil-plant conceptual reservoir model using a daily time step (Besbes et al., 2014). Both models were applied to reference domains in major cereal-producing regions. The green water-equivalent values provided by the hydrologic conceptual model are in accordance with the values generally cited in the literature. Based on the general assumption of uniform allowance for fertilizers amongst cereal-producing regions, these results indicate that cereal crop yields are directly related to rainfall inputs. These results are consistent with the formulation of the empirical model that produces, for the same regions, the same values of water-equivalents.

Green water assessment of rainfed agriculture

Figures 2 and A1 (in the online supplemental data at http://dx.doi.org/10.1080/07900627.2015.1040543) show the relation between annual rainfall and cereal and olive-tree yields, respectively. For olive farming, the best fit of the coefficient α is 0.25, meaning that a hydrological year of olive production contributes one-quarter in feeding olive trees while the preceding year counts for three-quarters. Figure 2 shows the correlation obtained for the most important cereals of Tunisia (durum wheat, wheat, barley and triticale). Focusing on total cereal production, the linear regression shows a relatively

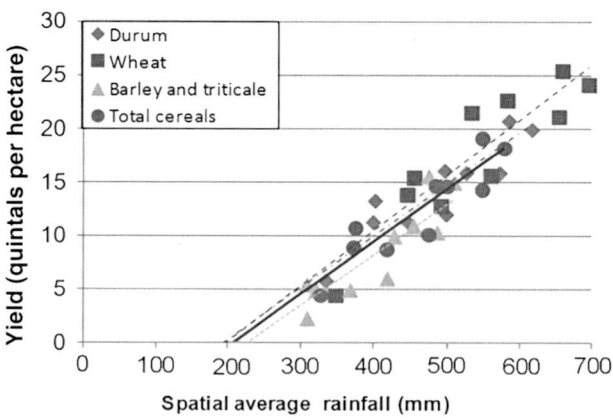

Figure 2. Cereal crop yields and spatial average annual rainfall, 2001–2010. Data from Besbes et al. (2014).

strong correlation. In the national average, mean annual rainfall weighted by sown area in the different regions is about 462 mm. As the rainfall threshold is about 210 mm, the water-equivalent of rainfed cereal crop production represents a rainfall of 252 mm. Applied to the average cultivated area (around 1,392,000 ha between 2001 and 2010), one may estimate the average cereals water-equivalent at almost 3.5 km^3 per year. Normalized to the national average production, this volume corresponds to around 2 m^3 per kg of cereals. Cereal production depends not only on rainfall but also on sown area, which is subject to significant variation. Applying the rainfall to the total area cultivated during 1996 (2.023 million ha), which corresponds to the maximum area sown between 1984 and 2010, one may estimate the potential of the Tunisian cereal agriculture at more than 5.1 km^3 water-equivalent per year.

The same analysis applied to rainfed olive trees indicates that the water-equivalent of olive production represents a rainfall of 108 mm. Applied to all Tunisian olive groves, whose total area is 1.666 million ha, one can evaluate the potential of green water associated with olive plantations at about 1.8 km^3. Applying the total green water volume to the total olive plantations, estimated at 65 million trees, the average specific volume is almost 22 m^3 per tree. There is no specific literature that would allow a systematic validation, but some comparisons can be made with Spanish data. Salmoral, Aldaya, and Chico (2010) estimated the green water equivalent of Spanish rainfed olive plantations at 1971 m^3/ha, which represents a water depth of 197 mm, nearly twice the value found for Tunisian plantations. As the Spanish olive groves are planted at relatively high density, typically 100 trees/ha (Galán et al., 2008), the green water specific volume stands at nearly 20 m^3/tree, very close to the value calculated for Tunisian olive trees.

On average, the potential associated with these two cultures is nearly 6.9 km^3, collected on a total area of 3.7 million ha, a specific volume of 1870 m^3/ha. Assuming that specific volume applies to all cultivated areas under rainfed conditions, whose potential is nearly 4.5 million ha (maximum rainfed-cultivated area between 1984 and 2010), one can estimate the potential of green water for rainfed agriculture to be about 8.5 km^3. Note that this resource drawn by rainfed agriculture from soil water is four times that of blue water resources allocated to irrigation. The most significant results related to the green water potential of rainfed agriculture are summarized in Table A1 (in the online supplemental data).

Water-equivalent of crop production

The data related to rainfed and irrigated agriculture serve to estimate the water-equivalent of vegetal production. Figure 3 shows the change in the water-equivalent of agricultural production. It appears that the water-equivalent of all crops represents an average of more than 9 km^3. However, this amount is subject to considerable variation, mainly due to rainfall and rainfed production variability; it varied from less than 2 km^3 in 2002 to over 11 km^3 in 2004. The irrigated production variability is naturally much lower, increasing from 1.1 km^3 in 1996 to over 1.5 km^3 in 2010, due to irrigated agriculture development.

Figure 4 shows the mean agricultural production water-equivalent structure for 2006–2010, where blue water represents 15% and green water 85%. This last green water part comes mainly from rainfed crops (75%) and partly from irrigated crops (10%). The latter percentage corresponds to the contribution of rainfall on the irrigated areas.

On average, irrigated agriculture water-equivalent represents only a quarter of the water resources involved in all agricultural production. In spite of its essential contribution, irrigated agriculture is too limited to be able to stabilize all agricultural production. Rainfed agriculture represents, on average, three-quarters of the agricultural

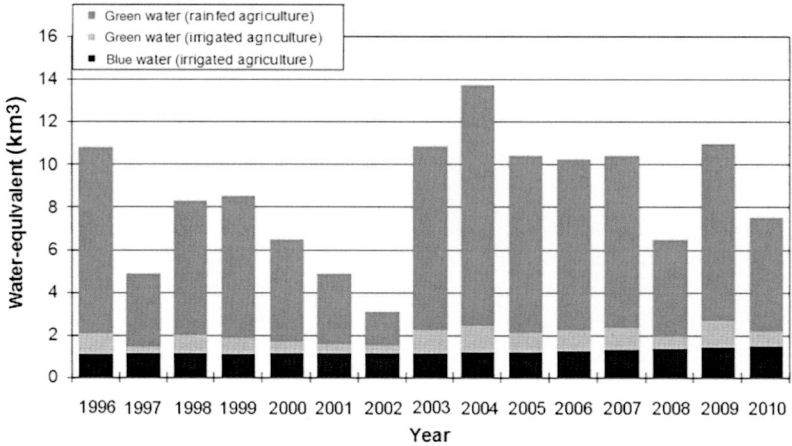

Figure 3.   Water-equivalent of vegetal production, 1996–2010. Data from Besbes et al. (2014).

production water-equivalent. However, rainfed agriculture is too variable and not able to satisfy all national food demand. As a consequence, Tunisia has an inbuilt need for additional contributions to fill the gap between local production and food demand, especially during periods of drought or low rainfall.

Water-equivalent of animal production

The animal production water-equivalent can be determined from the water-equivalent of products used for animal feed. In Tunisia, a significant part of fodder and grain used for animal feed is imported; another significant part comes from extensive grazing related to pastoral activity. Using the conversion factors of products intended for animal feed from the FAO (2006), one can rely on the precedent estimation for cereals water-equivalent, established at $2\,m^3/kg$, to estimate animal production water-equivalent.

The agri-food production of livestock origin has continuously increased during the last four decades, with a stabilizing trend during 2000–2010. Tunisia has reached self-

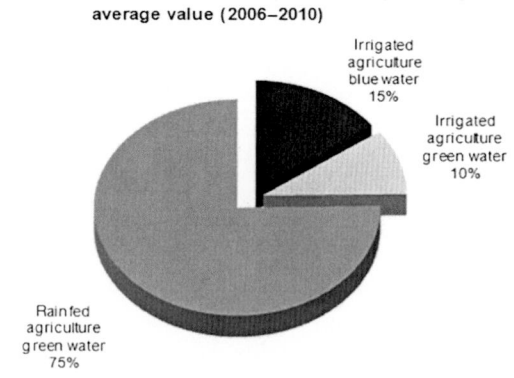

Figure 4.   Structure of the vegetal production water-equivalent, 1996–2010.

sufficiency in animal production through a sustained policy, including incentives for forage production. The change in the water-equivalent of animal origin products (meat, milk and eggs) is shown in Figure A2 (in the online supplemental data). This figure presents the different contributions to water-equivalent involved in various domains of livestock production over the last four decades. At the national scale, the agri-food production of livestock origin water-equivalent includes the water-equivalent of fodder produced locally from rainfed farming and irrigated crops (blue water and green water), the water-equivalent of fodder imports (virtual water) and the water-equivalent of natural rangelands (considered as green). Green water resources used in livestock production are variable: Deficits during low-rainfall periods are met by imports of grain or fodder (virtual water). The amount of virtual water is calculated by applying the water-equivalent obtained for domestic products to imported products.

Figure A3 (in the online supplemental data) shows the average water-equivalent of livestock production during 2006–2010. One can distinguish the quantities of blue water and green water associated with local production as well as the virtual water associated with imports of fodder and grain products intended for animal feed. It appears that the contribution of the irrigated sector remains weak: Local blue water is around 2% of the total animal production water-equivalent. Green water (rainfed fodder and rangeland production) provides a significant share of the fodder production. The rangelands account for over a quarter of the water balance of animal production. This represents the equivalent of nearly 1.7 million tonnes of fodder, more than twice the national average fodder production. This significant contribution is nevertheless provided by very large areas (4.7 million ha). It is known that the livestock sector is a very large anthropogenic user of land, and this is particularly true in arid regions.

In terms of water-equivalent, all the local forage production, including the contribution of rangelands, represents two-thirds of animal production. The sector of livestock production has an inbuilt need for an additional supply of fodder in the form of virtual water. The self-sufficiency in foodstuffs attained by the various animal production sectors during the last two decades is relative. From a water-equivalent point of view, the deficit of the sector amounts to about a third. Indeed, expressed in water-equivalent, it appears that the performance of the sector in achieving self-sufficiency in products of animal origin has been made possible by the contribution of fodder and cereal imports intended for animal feed.

Figure A4 (in the online supplemental data) shows the change in the water-equivalent of imports of agri-food products of livestock origin and fodder and cereals intended for animal feed. Encouragement of animal production during the late 1980s resulted in an increase in production and a decrease in imports of agri-food products of livestock origin. But this is offset by significant animal feed imports. In terms of water balance, this results in a significant increase of virtual water imports to support the sector of local livestock production. Within a decade these imports almost doubled, going from nearly 1 km^3 in the early 1990s to about 2 km^3 in the late 2000s; since then it seems that the imports have been stabilizing.

Water balance of food demand

Water-equivalent of food demand

The water-equivalent of all food production represents average water resources amounting to about 12 km^3/y. These come from water involved in rainfed and irrigated crop production, rangeland production and import-export balance of fodder and cereals employed in animal feed.

The change in water-equivalent of various forms of agricultural production indicates that the local water potential involved in agricultural production (blue and green water) has more than doubled over the past four decades, following the population increase from 5.1 million inhabitants in 1971 to 10.5 million in 2010. Tunisian agriculture succeeded in maintaining stable specific agricultural production (water-equivalent per capita) despite the significant increase in population (Table 1). On the other hand, Table 1 shows that irrigated agriculture made significant progress from 1971 to 2010: The blue water contribution increased from $0.58\,km^3$ to $1.36\,km^3$. It also appears that during the same period, the growth of food animal products was faster than that of crop production. Supported by fodder imports (reported as virtual water in Table 1), livestock production water-equivalent more than tripled. Correspondingly, the part of water resources involved in production of agri-food products of livestock origin significantly increased, from 28% to 41%. Consequently, foodstuffs of plant origin showed a decline of nearly 13%. During 2006–2010, the part of crop production amounts to 60% of the water-equivalent and the livestock production accounts for 40%. This change in the structure of food production also reflects a change in the structure of the Tunisian diet, in which the share of animal products has increased at the expense of vegetal foodstuffs (Figure A5 in the online supplemental data).

Figure A5 (in the online supplemental data) shows the change in the specific (per capita) foodstuffs demand, expressed in water-equivalent. The specific food demand has increased due to changes in population diet. It appears that water requirements for food have increased from about $1000\,m^3y$ per capita in the early 1970s to more than $1500\,m^3y$ per capita in the 2000s. One may also observe a significant increase in the water-equivalent related to products of animal origin linked with population diet change. Similar trends have been observed in developed countries (Renault, 2002). Building on the present level of Tunisian animal products consumption, one may expect more changes in diet, accompanied by increase in water requirements for food demand.

Concerning the origin and nature of water resources involved in food production during 1996–2010, Figure 5 indicates that the agricultural production issued from rainfed agriculture accounts for 57%, rangeland 11%, and irrigated agriculture 19% (12% in blue water and 7% in green water) of the water-equivalent of total agricultural production. Overall, the green water contribution (including rangelands) represents three-quarters of agricultural production. The remaining quarter corresponds half to the blue water associated with irrigation and half to the virtual water intake resulting from imports of fodder and grain intended for animal feed.

Water-equivalent of the food trade balance

The interpretation of food trade balance in terms of water-equivalent (virtual water) allows the assessment of the contribution of foodstuffs imports to the overall water balance of food demand. Local production varies due to the impact of climatic conditions on crop yields. These fluctuations determine the virtual water fluxes, where exports and imports fluctuate together with production. As a consequence, foodstuffs imports play an important part in filling the gap between food demand and local production.

Figure A6 (in the online supplemental data) shows the change in the water-equivalent of food imports and food production exports. The increase of virtual water imports has been significant in recent decades, growing from almost $2\,km^3/y$ to more than $8\,km^3/y$. During the same period, exports in terms of virtual water doubled, from about $1\,km^3/y$ to over $2\,km^3/y$. The food water-equivalent balance is in an increasing deficit. The coverage rate of

Table 1. Change in the average water-equivalent of foodstuffs production, 1971–2010. Data from Besbes et al. (2014).

Water-equivalent (million m³)		1971–1975	1975–1980	1981–1985	1986–1990	1991–1995	1996–2000	2001–2005	2006–2010
Animal products	Blue water	35	46	48	57	67	88	102	111
	Green water	1018	1332	1406	1661	1952	2543	2957	3226
	Virtual water	471	616	650	769	903	1176	1368	1492
	Total	1524	1994	2105	2487	2923	3807	4427	4829
Vegetal products	Blue water	544	589	664	735	792	914	1021	1249
	Green water	3417	3309	3682	3426	4909	5299	5294	5686
	Total	3961	3898	4346	4161	5700	6213	6315	6935
Food production	Blue water	579	635	712	792	859	1001	1123	1360
	Green water	4435	4641	5088	5088	6861	7842	8251	8912
	Virtual water	471	616	650	769	903	1176	1368	1492
	Total	5485	5892	6451	6648	8623	10,020	10,743	11,764

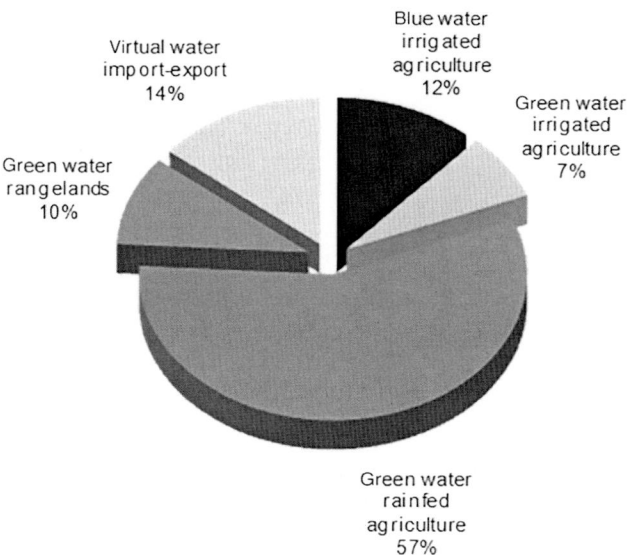

Structure of water-equivalent of foodstuffs production

Figure 5.   Structure of the water-equivalent of foodstuffs production, 1996–2010.

the food trade balance in water-equivalent (exports versus imports), which was 80% in the early 1970s, gradually deteriorated, until it stabilized at around 30% in the late 2000s.

In terms of volumes, the water deficit has widened significantly, from 560 million m^3 (100 m^3/y per capita) in the early 1970s to 6.16 billion m^3 (600 m^3/y per capita) in the late 2000s. As the growth in local food production has, very nearly, followed the growth of the population, the amplification of food imports and the consequent water deficit should be associated with the improvement and modification of the feeding structure, in particular in regard to the relative increase of foodstuffs of animal origin in diets.

The food trade balance analysis indicates that increase in food demand in Tunisia has resulted in a significant increase in the contribution of virtual water. Nevertheless, the food trade balance water-equivalent is very favourable to Tunisia: The coverage rate of the food trade balance which represents, in the late 2010s, 30% in terms of water-equivalent reached, in value, around 80% during the same period. This means that, by exchanging virtual water, the revenues from exports of food products (expressed in water-equivalent) are more important than the spending on imports of foodstuffs of an equivalent virtual water volume. The benefits of the commercial exchange of food products are due in particular to olive oil exports and to products which have an industrial added value such as grain products, fruits and canned food.

Conclusion

Assessments of water-equivalent related to food production and consumption show that the largest part of water resources is allocated to the agriculture sector, the largest part of food production is provided by rainfed agriculture, and a relatively large proportion of food demand is provided by foodstuffs imports. The water situation in Tunisia is such that it becomes increasingly difficult to continue the development of large-scale irrigated agriculture without consequences for the sustainability of water and soil resources. This perspective leads to the obvious conclusion that now, Tunisia has to rely on all of its resources to develop and support its agricultural production. Tunisia also needs to rationalize and optimize its virtual water fluxes.

The mean water-equivalent of all food production in Tunisia amounts to nearly 12 km^3/y. This resource comes from water involved in the production of rainfed and irrigated crops, including rangelands, and the water-equivalent of imports of fodder and grain intended for animal feed. Crop production amounts to nearly 60% of the water-equivalent for food production; livestock production accounts for about 40%. Food production from rainfed agriculture accounts for 56% of the water-equivalent of agri-food production, rangelands around 11%, and irrigated agriculture 19% (12% from blue water and 7% from green water) of all food production water-equivalent.

Overall, the contribution of green water, including rangelands, represents three-quarters of food production. The remaining quarter represents about half of the virtual water intake resulting from imports of fodder and grain for livestock production; and for the other half, the blue water associated with irrigation. Therefore, it appears that all blue water resources allocated to the irrigation sector, which represents 80% of mobilized water resources, contribute 12% of the water-equivalent of local agricultural production. The remaining part is directly provided by rain. Therefore, enhancing the potential growth of green water remains a key instrument in fulfilling food security objectives.

While the population has almost doubled over the past four decades, the water-equivalent of food demand has more than tripled. This means that the per capita water-equivalent of food demand rose by more than half in that time period. The total water-equivalent of food demand amounted to 16.4 km^3 in 2010. To meet this demand, Tunisia imports 8.3 km^3 of virtual water in the form of foodstuffs (cereals, fodder, vegetable oil, sugar, etc.). At the same time, it exports agricultural products (citrus, dates, vegetables, olive oil, etc.) equivalent to 2.5 km^3/y, leaving a balance of 5.8 km^3/y (over a third of total food needs). The net contribution of virtual water is partly (4.3 km^3/y) associated with balance of trade balance of foodstuffs; and the rest (1.5 km^3/y) corresponds to the contributions in virtual water associated with fodder imports to support local livestock production.

This component split of water-equivalent related to Tunisian food demand indicates that the water-security issue has to be examined from the angle of its connection with the critical issue of food security, its relationship to agricultural production, and food trade balance. The water-equivalent balance of the agro-food trade is in deficit, and this deficit is in danger of growing further. To alleviate its impacts, Tunisia has to redouble its efforts to improve the performance of both irrigated and rainfed production and to rationalize virtual water fluxes by managing food demands, supplies and reserves. Integration of all water resources at the national scale is essential in facing the great challenges of food security. This is particularly true in semi-arid countries where, as in Tunisia, rainfed agriculture plays a key role in food security. Rainfed agriculture should be regarded as a mode of development of water and soil resources for which appropriate solutions have to be developed. The main objective is to make rainfed farming viable and profitable in order to further expand its production and consolidate its economic and social roles. These objectives require structural solutions to enhance its contribution (extent of rainfed farming, development of foodstuffs stockpiling, etc.) and non-structural solutions to improve the performance of the sector (trade, drought management, insurance, etc.).

The results obtained through this research give pertinent orders of magnitude of the available water resources for agricultural production. However, these initial results need to be interpreted carefully and need to take into consideration the limitations of the quantitative approach used in this study. Indeed, the linear regression model cannot fully reflect all the complexities of agricultural production, which involves non-water-related agronomic practices (soil management techniques, fertilization, effective control of weeds) that play important roles in increasing crop water productivity. These and all other external factors need to be considered, in order to include, in a holistic approach, all the considerations involved in the development and management of water and soil resources.

Supplemental data

Supplemental data for this article are available at http://dx.doi.org/10.1080/07900627.2015.1040543

Disclosure statement

No potential conflict of interest was reported by the authors.

References

Aldaya, M. M., Allan, J. A., & Hoekstra, A. Y. (2010). Strategic importance of green water in international crop trade. *Ecological Economics, 69*, 887–894. doi:10.1016/j.ecolecon.2009.11.001

Besada, H., & Werner, K. (2015). An assessment of the effects of Africa's water crisis on food security and management. *International Journal of Water Resources Development, 31*, 120–133. doi:10.1080/07900627.2014.905124

Besbes, M., Chahed, J., & Hamdane, A. (2014). *Sécurité hydrique de la Tunisie, Gérer l'eau en conditions de pénurie*. L'Harmattan.

Besbes, M., Chahed, J., Hamdane, A., & De Marsily, G. (2010). Changing water resources and food supply in arid zones: Tunisia. In G. Schneider-Madanes & M. F. Courel (Eds.), *Water and sustainability in arid regions*. dordrecht: Springer.

Chahed, J., Besbes, M., & Hamdane, A. (2011). Alleviating water scarcity by optimizing 'Green Virtual-water'. In A. Y. Hoekstra, M. M. Aldaya, & B. Avril (Eds.), Value of Water Research Report Series, No. 54, UNESCO-IHE, Delft, the Netherlands (pp. 99–114).

Chahed, J., Hamdane, A., & Besbes, M. (2008). A comprehensive water balance of Tunisia: Blue water, green water and virtual water. *Water International, 33*, 415–424. doi:10.1080/02508060802543105

Doorenbos, J., & Kassam, A. H. (1979). Yield response to water. FAO Irrigation and Drainage Paper No. 33. Rome, FAO.

Falkenmark, M. (2001). The greatest water problem: The inability to link environmental security, water security and food security. *International Journal of Water Resources Development, 17*, 539–554. doi:10.1080/07900620120094073

FAO. (2006). Livestock's long shadow, environmental issues and options. FAO, Rome 2006.

FAO. (2014). AQUASTAT, FAO's global information system on water and agriculture. FAO, http://www.fao.org/nr/aquastat, last access: 03October 2014, Rome, Italy.

Galán, C., García-Mozo, H., Vázquez, L., Ruiz, L., Díaz de la Guardia, C., & Domínguez-Vilches, E. (2008). Modeling olive crop yield in Andalusia. *Spain Agronomy Journal, 100*, 98–104.

Gargouri, K., Rhouma, A., Sahnoun, A., Ghribi, M., Ben Taher, H., Ben Rouina, B., & Ghrab, M. (2008). Assessment of the impact of climate change on olive growing in Tunisia using GIS tools. *Options méditerranéennes, Series A N°80*.

Ghrab, M., Gargouri, K., & Ben Mimoun, M. (2008). Long-term effect of dry conditions and drought on fruit trees yield in dryland areas of Tunisia. *Options méditerranénnes, Series A N°80*.

Hamdane, A. (2007). Tunisia National report on monitoring progress and promotion of water demand management policies. Water demand management in the Mediterranean, progress and policies. Blue Plan, Zaragoza, Spain.

Hoekstra, A. Y., & Chapagain, A. K. (2006). Water footprints of nations: Water use by people as a function of their consumption pattern. *Water Resources Management, 21*, 35–48. doi:10.1007/s11269-006-9039-x

Hoekstra, A. Y., & Mekonnen, M. (2011). The water footprint of humanity. In Peter H. Gleick (Ed.), *Pacific Institute for studies in development, environment and security*. www.pnas.org/cgi/doi/10.1073/pnas.1109936109

MARH. (1998–2007). Annuaires des statistiques agricoles. Ministère de l'Agriculture et des Ressources Hydrauliques, Tunisia.

Mekonnen, M. M., & Hoekstra, A. Y. (2011). The green, blue and grey water footprint of crops and derived crop products. *Hydrology and Earth System Sciences*, *15*, 1577–1600. doi:10.5194/hess-15-1577-2011

Redwood, M., Bouraoui, M., & Houmane, B. (2014). Rainwater and greywater harvesting for urban food security in La Soukra, Tunisia. *International Journal of Water Resources Development*, *30*, 293–307. doi:10.1080/07900627.2013.837367

Renault, D. (2002). *Value of virtual water in food: Principles and virtues* Workshop on Virtual Water Trade, 12-13 December 2002. Delft, Value of Water Research Report Series, No. 12, UNESCO-IHE, Delft, the Netherlands, Ed. Hoekstra (pp. 70–91).

Renault, D., & Wallender, W. W. (2000). Nutritional water productivity and diets. *Agricultural Water Management*, *45*, 275–296. doi:10.1016/S0378-3774(99)00107-9

Rezgui, M., Ben Mechlia, N., Bizid, E., Kalboussi, R., & Hayouni, R. (2005). Étude de la stabilité du rendement de blé dur dans différentes regions de la Tunisie. In: L'amélioration du blé dur dans la région méditerranéenne: nouveaux défis. Options méditerranéennes. *Série A: Séminaires Méditerranéens 2000, 40*, 167–172.

Salmoral, G., Aldaya, M. M., & Chico, D. (2010). The water footprint of olive oil in Spain. Fundación Marcelino Botín Eds.

Steduto, P., Hsiao, T. C., Fereres, E., & Raes, D. (2012). Crop yield response to water. FAO Irrigation and Drainage Paper No. 33. Rome.

Zimmer, D. (2013). L'empreinte-eau. Mayer Charles Leopold Eds.

Energy cost of irrigation policy in Morocco: a social accounting matrix assessment

Mohammed Rachid Doukkali[a] and Caroline Lejars[a,b]

[a]Social Science Department, Institut National Agronomique Hassan II (IAV Hassan II), Al Irfane, Rabat, Morocco; [b]UMR-GEAU, Centre de Coopération International en Recherche Agronomique pour le Développement, Montpellier, France

The objective of this study was to assess the consumption and the multiplier effect of the use of energy and irrigation water for rainfed and irrigated agriculture at the national level in Morocco. Using a social accounting matrix, the direct and indirect economic effects of subsidizing energy used by agriculture were identified. The results show that irrigation water policy in Morocco, which targets 'water-saving' techniques, has increased the use of subsidized energy and that indirect effects, through energy subsidies, exceed the direct effects of agricultural subsidies. A social accounting matrix can help decision makers make the necessary trade-offs between irrigated and rainfed agriculture.

Introduction

The demand for water, food and energy is expected to increase by 30–50% in the next two decades (WEF (World Economic Forum), 2011). The world's water and energy resources are already experiencing significant stress and shortfalls (Bazilian et al., 2012; Smil, 2000; Waughray, 2011), yet rapidly increasing demand for these resources is foreseen in the coming years (Hoff, 2011). In the literature, this anticipated increase is specifically referred to in the context of growing populations, increasing rates of urbanization, expanding middle-class lifestyles and diets, and an overall increase in the demand for resources (Hoff, 2011; Van Vuuren et al., 2012; WEF (World Economic Forum), 2011).

While the interconnected nature of water, energy and food (the WEF nexus) has been recognized and confirmed by several studies around the globe (Granit et al., 2012; Malik, 2002; Scott & Shah, 2004), there is a relatively limited understanding of how to tackle these complex relationships though assessments and actions. As underlined by the debate at the WEF (World Economic Forum) (2011), any strategy that focuses on one component of the WEF nexus without considering its interconnections risks serious unintended consequences. It is necessary to improve policy coordination and harmonization to account for trade-offs and to build on the increased interconnectedness of water, energy and food. Part of this process is promoting, identifying, and eliminating contradictory policies (WEF (World Economic Forum), 2011).

In Morocco, as in many developing countries (Besada & Werner, 2015), the WEF nexus is a critical issue. The country has to reconcile four overlapping conditions: heavy reliance on imported energy; a growing population; recurrent drought and groundwater overexploitation; and the need to increase economic growth. In this context, the agriculture sector is expected to play a key role in improving the country's overall economic growth and earnings from exports, as well as in improving food security and poverty alleviation. Public agricultural policies mainly aim to achieve food security and increase exports (MEF (Ministère de l'économie et des finances), 2012). These policies are based on the intensification of production. They require considerable investment and natural resources, and rely to a large extent on irrigation. Despite higher energy consumption and rising energy prices, agricultural policies and water policies are still disconnected from energy policies, which may result in conflicting objectives and inconsistent development policies.

A social accounting matrix (SAM) is an appropriate framework with which to identify and assess the effects of inconsistencies between policies. Basically, it is an extension of the national accounting system that provides a conceptual basis for examining both growth and distribution issues in the economy within a single analytical framework. Since an SAM represents the whole economic system, it highlights the linkages between the different accounting components of the system such as commodities, activities, factors and institution accounts, as well as foreign accounts. It is commonly used in the general analysis of agricultural and rural development policies (FAO, 2012), as well as in analysis of specific components of these policies such as irrigation water policies (Doukkali, Flichman, & Faurès, 2004). An SAM could be also used to study the linkages between energy, water and food policies, and to assess the multiplier effects related to the use of energy and water in agriculture.

Classically, and in line with the traditional central economic planning techniques developed in the first half of the twentieth century (known as Leontief input-output analysis), the SAM and its derived multipliers are used to measure the impact on the local economy of an increase in one or in several components of an exogenous demand, such as government, investment, or export demand (see e.g. Miller & Blair, 2009). As one objective of this study was to assess the impact of the ongoing agricultural and energy policies on the local economy and to measure direct and indirect use of energy implied by the agriculture production process, a new structure of SAM was developed and specific production multipliers were estimated. These multipliers measure the use and consumption of goods and services directly or indirectly involved in the production of agricultural output, particularly in terms of energy.

This article aims (1) to demonstrate the usefulness of using an SAM and production multipliers to analyze the consistency of water, energy and agricultural policies, and (2) to apply this approach to the case of Morocco. First, we describe current energy, irrigation and agricultural policies in Morocco. Second, we present the SAM and the production multipliers. Finally, we assess the multiplier effects that capture the backward and forward linkages between energy, water and food policies at the national level in Morocco. Using this SAM and these multipliers, it is possible to determine the economic effects of energy and irrigation subsidies on rainfed and irrigated agriculture.

Case study: irrigation, energy and agricultural policies in Morocco

Agriculture, and particularly irrigation, has been a major component of the development policy pursued by Morocco during the last half-century. Given its semi-arid climate and erratic rainfall (inter- and intra-annually), Morocco had to invest heavily in irrigation to prevent climatic risks, to ensure minimum food security, to obtain rapid growth of land productivity and to promote agricultural exports. This strategy was also explained by the fact that the agriculture sector is the main source of income for more than 45% of the population, which is still rural, and provides jobs for 40% of the employed active population.

Currently, with 1.46 million hectares of irrigated land (15.8% of the country's arable land), irrigation contributes to around 45% of the agricultural added value in a normal year (World Bank, 2013) and accounts for around 85% of total freshwater used. Irrigation benefits 36% of 1.5 million farms, and of the 36%, 72% are smallholdings (see Table A1 in the online supplemental data at http://dx.doi.org/10.1080/07900627.2015.1036966). Irrigation systems in Morocco are mainly based on gravity (80.8% of irrigated land), and localized irrigation represents only 9.7%.

In the last five years, after more than two decades of structural adjustment and liberalization of the agricultural sector, with the collapse of the Doha negotiations and the spikes in world agricultural prices in 2007/2008, Morocco has embarked on an ambitious and aggressive agricultural strategy. This strategy is mainly based on public financial support to private investment and an impressive increase in public spending and subsidies to the sector, in particular to irrigation (Figure A1 in the online supplemental data). The aim of this new strategy, which was launched in 2008 and is called the Green Morocco Plan (GMP), is to implement a vast array of structural and sectorial reforms in agriculture. The GMP's strategic objective is to strengthen the sector's competitiveness while stimulating massive private investment in agriculture, estimated at MAD 10 billion per year (MAPM (Ministère de l'Agriculture et de la Pêche Maritime), 2012), to accelerate growth of the agricultural sector and to increase exports. The GMP promotes high-value-added crops (fruit trees and vegetables) and intensive livestock production (dairy and red meat) by subsidizing private investment in general and irrigation in particular. In addition to an increase in the Ministry of Agriculture's investment budget (Figure A1 in the online supplemental data), the subsidy fund for agriculture (the Agriculture Development Fund) increased by 600% between 2007 and 2012. In 2011, total direct subsidies to agricultural private investment from the FDA amounted to MAD 2.3 billion. In accordance with the GMP, the Moroccan government also adopted a new long-term irrigation program called the National Irrigation Water Saving Program Support Project, which was adopted in December 2009. The project has two aims: to convert 550,000 ha of land already irrigated by gravity and sprinkler to localized irrigation; and to convert existing cropping systems to high-value-added crops.

This new strategy, whose main aims are to increase exports and achieve food security, led to the rapid intensification of agricultural production, which requires more energy per hectare. These policies led to farming practices and cropping systems that require both considerable investment and natural resources, and that rely on irrigation, and thus require more energy, especially for pumping. Consequently, between 2004 and 2011, according to the Ministry of Agriculture and the Department of Mining and Energy, the consumption of energy per hectare increased by 40% (from 120 toe/kha to 182 toe/kha).

The Green Morocco Plan was designed without taking the implicit increase in the demand for energy or the bill for imported energy into account. Currently, the country is heavily dependent on external sources of energy and is vulnerable to volatile oil, gas and coal prices. Morocco imports 96% of its energy, and its energy needs are increasing by an average of 8% per year. In 2011, the cost of these imports amounted to more than MAD 100 billion, which corresponded to one-fourth of the country's total imports (Figure A2 in the online supplemental data). According to the Department of Energy and Mining, 13% of national energy is consumed by the agricultural sector (Table 1), mainly for irrigation, machinery and livestock. In 2011, 46% of the energy used by the agricultural sector was LPG, and 45.5% was diesel, while electricity accounted for only 7.9%.

Table 1. Energy final consumption in 2011 according to origin and sector of use, in KTOE.

Sector	Petroleum products final consumption					Electricity final consumption	Total	Per cent
	Diesel	Gasoline	Gas	Others	Subtotal			
Household	19	0	1,326	6	1,351	724	2,075	16.5
Agriculture	765	10	774	0	1,549	132	1,681	13.4
Transport	4,079	620	0	600	5,299	26	5,325	42.4
Industry	71	−25	197	1,907	2,150	1,326	3,476	27.7
Total in KTOE	4,934	605	2,297	2,513	10,349	2,208	12,557	100.0
Total in per cent	39.3	4.8	18.3	20.0		17.6	100.0	

Note: KTOE, Kilotonne of Oil Equivalent.

Moreover, energy products are highly subsidized in Morocco, and the subsidy system not only distorts product prices but also places a heavy burden on the government budget. In fact, in 2011, subsidies for petroleum products amounted to MAD 44.5 billion, i.e. 5.4% of the GDP, and the equivalent of 89% of the total public investment budget. In many cases, the subsidy represents a high percentage of the sale prices, e.g. 220% for butane (World Bank, 2013)

Despite increased energy consumption and the high level of subsidy, agricultural policies and water policies are still designed independently of energy policies. This could result in an increase in Morocco's dependency on energy imports. There is thus an urgent need to analyze the linkages between energy, water and food policies and to assess the multiplier effects of the use of energy and water for agriculture. In this context, the SAM could be a useful tool to identify and assess the effects of possibly conflicting policies.

Methodology: building a new SAM and estimating production multipliers

The SAM technique is related to national income accounting that represents the whole economic system. SAM is an established technique for capturing the details of disaggregated national accounts, in which the data are displayed in single-entry matrix format, rather than in the traditional form of double-entry bookkeeping (System of National Accounts, 1993). An SAM can serve as a statement of the initial conditions in an economy and as a starting point for a theoretical analysis of the mechanics of growth or the likely effects of policies. Different SAMs have been constructed for the Moroccan economy, including the one applied in a computable equilibrium model developed by Diao, Doukkali, and Bingxin (2008) and updated by Doukkali (2012). However, some of the data used by these SAMs are less appropriate for the issue at hand, since they did not disaggregate the energy subsectors. For this reason, a new SAM was constructed that takes the needs of the WEF nexus analysis into account. This new SAM is based on data from the Department of Energy and the Ministry of Agriculture. These data provide information on the total area of land under cultivation, changes in cultivated and irrigated areas per agricultural subsector, and energy consumption per economic sector and type of energy.

The newly constructed SAM

Several studies and data were used to build a new disaggregated SAM for 2011.

First, a supply-use table and institutional sector accounts were needed to compile a national accounting matrix. We used the 2011 supply-use table from the national accounts (Moroccan High Commission for Planning, 2013), which is disaggregated into 20 sectors and subsectors – as production accounts in the Activities sub-matrix, and as goods and services accounts in the Commodities sub-matrix. In this supply-use table, the agricultural sector is represented by a single account (A00), while energy is split into three: extractive industries (C00), oil refining and other energy production (D23), and electricity and water (E00).

Next, the agriculture and energy accounts were disaggregated to better understand the linkages between water, energy and food. Given that energy consumption differs between the irrigated and non-irrigated sectors, we further divided each agricultural sector into 'irrigated' and 'rainfed'. The electricity and water accounts were also disaggregated to separate the water subsector from the electricity subsector. The electricity sub-branch was also disaggregated according to the type of energy used to produce it (water, wind, or solar). Given that Morocco is almost completely dependent on imports of fossil fuels, the mining industries and oil refining sectors were kept aggregated in the activities accounts. In terms of goods and services, the three sectors including energy (C00, D23 and E00) were disaggregated to distinguish the different energy products as separate commodities, to assess the taxes and subsidies embedded in the products, and consumption by end users.

The agriculture account was disaggregated into 10 sub-production accounts, including four groups of rainfed crops (cereals and pulses, fodder crops, industrial crops and fruit trees), five groups of irrigated crops (the same as those in the rainfed agriculture group, plus vegetables) and a final account for 'other activities' including other crops, livestock, forestry and ancillary services to agriculture. The 2009 agricultural economic accounts were updated to 2011 using the annual production and price statistics published by the Ministry of Agriculture.

The three sectors that produce energy (C00, D23 and E00) were disaggregated based on the disaggregation of energy accounts produced by the Moroccan High Commission for Planning in 2007. The economics accounts for energy were updated to 2011 using the Moroccan energy balance sheet published by the Ministry of Mining and Energy in 2011, the price statistics published in the Statistical Yearbook, and data on taxes and energy subsidies published by the Ministry of Finance.

In total, the SAM for 2011 used in this article has 77 accounts: 33 production activities accounts, 30 commodities accounts, 2 primary factors accounts (labour and capital), 2 institutional accounts (households and government budget), 7 taxes and subsidies accounts, 1 savings-investment account, 1 inventories variations account, and the rest of the world. See Figure 1 for the accounts retained and Figure A3 (in the online supplemental data) for an aggregated version of the SAM.

Multipliers of production: calculation and interpretation

Formerly, multipliers were computed from an input-output table where the final demand for all its components (household demand, investment and exports) was considered to be exogenous (Leontief, 1986). The multipliers were used to calculate the backward linkages in the economy and to estimate the production required by each sector (the endogenous variable) to satisfy the increase in demand (the exogenous variable). With the development of the SAM, the multipliers were used to identify not only backward but also forward linkages (Pyatt & Round, 1985). In this way, the SAM multipliers can capture the total direct and indirect effects in the first and all subsequent rounds of the circular income flow. In most published studies, SAM multipliers have been used to measure the impact on the local economy of an increase in what most authors considered an exogenous demand (government, investment, or/and export demand; Bell, Hazell, & Slade, 1982; Defourny & Thorbecke, 1984; Pyatt & Round, 1985). More specifically, multipliers are used to evaluate the impact of changes in exogenous demand (for example, increased demand for agricultural exports) on exogenous variables (changes in total production and income). More recent studies have provided an alternative detailed way to break down these multipliers (Miller & Blair, 2009; Yang, Thurlow, & Lahr, 2012).

Since the objective of this study was to measure the impact of agricultural production on the local economy, production-specific multipliers were calculated to measure the use and consumption of goods and services directly or indirectly involved in the production process, particularly in terms of energy. To compute these multipliers, the only exogenous accounts used were production accounts, in other words, the activity-commodity (AC) sub-matrix, as shown in Figure 2.

Activities (33)	Commodities (30)	Factors and Institutions (14)
Rainfed Cereals and Pulses	Agriculture, Forestry and related services	Gross Wages
Irrigated Cereals and Pulses	Fishery products	Gross Operating Surplus/Goss Mixed Income
Rainfed Fodders	Coal	Households
Irrigated Fodders	Crude Oil	Government Budget
Rainfed Industrial Crops	Natural Gas	Taxes on production
Irrigated Industrial Crops	Other Products of the Mining Industry	Production subsidies
Irrigated Vegetable Crops	Food and Tobacco Industries	Tariffs and taxes on imports
Fruit Tree Crops	Textile and Leather	Value added taxes
Fruit Tree Crops	Chemical and Para-chemical Industries	Taxes on Commodities
Other Agriculture, Forestry and Auxiliary Services	Mechanical, Metallurgical and Electrical Industries	Subsidies to Commodities
Fisheries, Aquaculture	Manufacturing and Other Industries (excluding oil refining)	Direct Taxes (income taxes)
Mining Industries	Gasoline	Saving / Gross Fixed Capital Formation
Food and Tobacco Industries	Diesel	Inventories Variations
Textile and Leather	Butane and Propane	Rest of the World
Chemical and Para-chemical Industries	Other Petroleum Products	
Mechanical, Metallurgical and Electrical Industries	Thermal Electric Power	
Manufacturing and Other Industries.(excluding oil refining)	Hydro-Electricity	
Refining of Oil and Other Energy Products	Wind Electricity	
Thermal Electric Power	Solar Electricity	
Hydro-Electricity	Water	
Wind Electricity	Construction and Public Works	
Solar Electricity	Commerce	
Water	Hotels and Restaurants	
Construction and Public Works	Transportation	
Commerce	Post and Telecommunications	
Hotels and Restaurants	Financial and Insurance Activities	
Transportation	Real Estate, Renting and Serv. Made Companies	
Post and Telecommunications	Public Administration and Social Security	
Financial and Insurance Activities	Education, Health and Social Work	
Real Estate, Renting and Serv. Made Companies	Other Non-Financial Services	
Public Administration and Social Security		
Education, Health and Social Work		
Other Non-Financial Services		

Figure 1. The seventy-seven accounts retained after disaggregating the 2011 social accounting matrix.

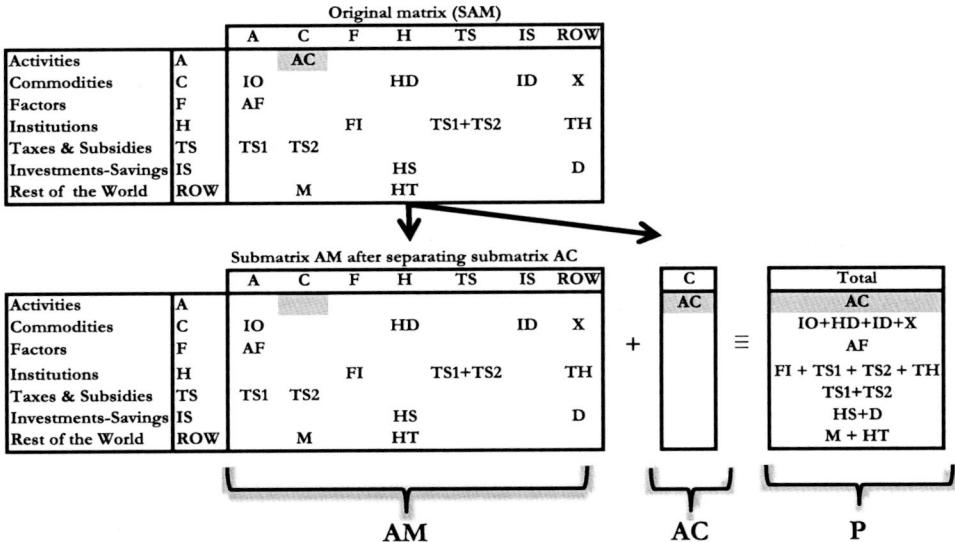

Figure 2. Original social accounting matrix broken down into two sub-matrices.
Note: The taxes and subsidies submatrix (TS) was divided in two submatrices (TS1 + TS2), so as to separate taxes and subsidies linked to energy and others taxes and subisidies.

To compute these multipliers, the column vector of total rows of the SAM (P) was broken down into the sum of two columns of vectors. The first is the vector AM, which is the sum of each row of the SAM without the sum of the row of the sub-matrix that gives the transformation of the production of domestic commodities into domestic outputs, also called the activity-commodity matrix (AC), and the second is the vector of the sums of the rows of the sub-matrix AC. That is:

$$P \equiv AM + AC. \tag{1}$$

Dividing each component of the sub-matrix AM by the total of the corresponding column of the SAM, the sub-matrix AM can be written as

$$AM \equiv MTC \times P \tag{2}$$

where MTC is the sub-matrix of the SAM technical coefficients, excluding those corresponding to the sub-matrix AC, and $\times$ is the vector product.

Writing the vector P in a matrix format as the product of the unity matrix I multiplied by the column vector P itself, and rearranging the terms in Equations (1) and (2), gives:

$$P = (I - MTC)^{-1} \times AC = MPM \times AC \tag{3}$$

In Equation (3), the matrix MPM, which corresponds to the inverse of the matrix I − MTC, is the SAM's matrix of production activities multipliers. The column of a given activity in the matrix MPM gives, for each unit produced domestically by that activity: (i) how much directly and indirectly incorporated (or footprint) intermediate inputs it requires, including energy; (ii) how much primary factor income (labour and capital) or institutional revenues (households and government budget), it generates directly or indirectly; and (iii) how many imports, investments, taxes, or subsidies this involves, directly or indirectly.

Results: assessing budgetary costs and the multiplier effects of water and energy policies

Energy consumption in the Moroccan economy

The constructed 2011 SAM (Figure A3 in the online supplemental data) reveals that the Moroccan GDP is still low in comparison with the size of the population (approximately 33 million). Per capita, this represents an average of MAD 28,000, placing Morocco among the countries with low average incomes. Although it is a small economy with a low GDP, the Moroccan economy is relatively open to the international market. Total imports of goods and services in 2011 amounted to MAD 400 billion, representing around 50% of GDP. The goods and services trade deficit was approximately 30%.

Energy imports contribute substantially to this deficit, since the net energy import bill represents 21% of total imports. Energy use in the economy not only contributes to the trade deficit but also increases the budget deficit. In 2011, total subsidies to energy represented more than 5% of total GDP. Import taxes on energy imports represented 2.5% of total GDP. Finally, total energy consumption, at the domestic market price, represented MAD 103 billion, and energy used directly by agriculture MAD 11 billion, i.e. 11.9% of total energy consumption.

Budgetary cost of energy subsidies

The SAM shows that the agricultural sector consumes (in MAD) 9.88% of energy, i.e. 11.4% of the total consumption of petroleum products and 5% of total electricity consumption.

In 2011, total energy subsidies reached MAD 44.3 billion (Table 2). Gasoline accounted for 4.2% of total subsidies, diesel for 37.7%, gas for 30.9%, and the rest for 27.2%. The taxes generated by the use of energy amounted to MAD 22.16 billion, or 50% of the total energy subsidies. For 2011, the SAM estimates indirect energy subsidies to the agricultural sector at MAD 5.26 billion. Indirect subsidies for energy thus greatly exceeded direct subsidies to the agricultural sector (MAD 2.3 billion for the Agricultural Investment Fund in 2011).

In agriculture, butane is used only for private irrigation. Consequently, we estimate that MAD 3 billion were allocated to private irrigation through butane subsidies alone, though this subsidy was supposed to benefit only energy use by households. In practice, the 'water-saving' policy, targeting 'water-saving' techniques, has led to increased use of subsidized butane in agriculture. These subsidies for butane benefit the majority of smallholders (73% of irrigated farms have less than 5 ha).

Table 2. Budgetary costs of direct energy subsidies (MAD millions).

Energy products	Total subsidies	Direct subsidies to energy in agriculture
Gasoline	1,878	29
Fuel oil	16,695	2,112
Gas (butane and propane)	13,715	3,037
Other oil products	12,050	–
Total subsidies to gas and oil products	44,338	5,178
Subsidies through electricity		83
Total energy subsidies	44,338	5,260

Investment cost of the reconversion to localized irrigation and the expansion of large-scale irrigation

The GMP foresees the reconversion of 555,000 ha between 2008 and 2020. In 2007, the budget for reconversion to localized irrigation was estimated at MAD 37 billion and was supposed to save $14\,\mathrm{km}^3$ of water per year. However, these calculations were made assuming that 60% of the reconversion program would be financed by the state and 40% by the beneficiaries. Until the beginning of the reconversion program in 2008, nearly 100% had been financed by the state. Consequently, the effective amount of the subsidy for reconversion was higher than expected and can be estimated at MAD 61 billion. The budgetary cost projected to support reconversion is thus MAD 111,000 per reconverted hectare.

In terms of irrigation expansion, 108,440 additional hectares should also be newly equipped by 2020. The budget allocated to cover this expansion is MAD 15 billion, and includes the participation of farmers in the purchase of equipment. The budgetary cost projected to support extension is thus MAD 138,440 per irrigated hectare of expansion.

The high budgetary investment in irrigation for agriculture will affect only 16% of total arable land and potentially the 36% of the country's farmers who already have access to irrigation, 73% of whom farm less than 5 ha. In other words, these subsidies, which were designed for reconversion to localized irrigation, benefit only the one-third of farmers who already have access to irrigation water. Agricultural intensification calls for increasing investments in energy and irrigation.

Multiplier effects

Table 3 lists the SAM multipliers for the six agricultural subsectors, for rainfed and irrigated crops separately.

Regarding commodities, the multipliers of the irrigated subsectors are higher than the multipliers of the rainfed subsectors for all crops. This is logical because, irrespective of the type of crop, irrigation consumes more inputs per unit of outputs. But the multiplier effects of final energy consumption are significantly higher for irrigated crops than for rainfed crops, whereas the multiplier effects of other goods and services are lower for irrigated than for rainfed crops.

Regarding primary revenues (labour and capital), except for irrigated fodder crops, the multipliers are higher in the rainfed sector than in the irrigated sector. Indeed, irrigated fodder crops have an effect on livestock, which is an income-generating factor. Production in the rainfed sector also produces more income for households and for the government than production in the irrigated sector (except for irrigated fodder crops).

With respect to taxes, the multipliers are slightly higher in the irrigated sector. This is consistent with the fact that taxes apply mainly to goods and services rather than to products. However, the resulting net taxes (i.e. taxes minus subsidies) show that the multiplier effects are higher for rainfed than for irrigated crops. Finally, as the rest-of-the-world multipliers are higher for irrigated than for rainfed crops, it is clear that the rainfed sector requires far fewer imports than the irrigated sector.

As a result, and based on analysis of multiplier effects, in terms of value added and revenues per MAD invested, investing in rainfed agriculture could be as profitable as investing in irrigated agriculture. Even if irrigation is an important component of agricultural production in Morocco, the current agricultural policy should be more balanced in favour of rainfed agriculture, and the government should now promote more rainfed rather than irrigated crops.

Table 3. Matrix production multipliers for the social accounting matrix.

	Cereals and pulses		Fodder crops		Industrial crops		Fruit tree crops		Vegetables	Other agriculture
	Rainfed	Irrigated	Rainfed	Irrigated	Rainfed	Irrigated	Rainfed	Irrigated	Irrigated	
Activities	1.00	1.00	1.00	1.00	1.00	1.00	1.00	1.00	1.00	1.00
Commodities	1.27	1.33	1.26	1.30	1.27	1.29	1.27	1.32	1.28	1.29
Final energy consumption	0.07	0.30	0.07	0.22	0.07	0.15	0.07	0.22	0.13	0.12
Other goods and services	1.20	1.03	1.19	1.09	1.20	1.15	1.20	1.10	1.16	1.17
Primary factors	0.73	0.44	0.70	0.73	0.99	0.79	0.96	0.71	0.64	0.68
Households	0.96	0.60	0.92	0.95	1.28	1.04	1.24	0.94	0.85	0.91
Government revenue	0.20	0.06	0.20	0.16	0.26	0.20	0.26	0.16	0.16	0.19
Taxes	0.13	0.14	0.12	0.15	0.15	0.14	0.15	0.15	0.13	0.14
Subsidies	−0.03	−0.14	−0.03	−0.09	−0.03	−0.06	−0.03	−0.10	−0.06	−0.06
Taxes minus subsidies	0.10	0.00	0.09	0.06	0.12	0.08	0.12	0.05	0.07	0.08
Rest of the world	0.27	0.39	0.26	0.35	0.28	0.32	0.29	0.37	0.31	0.30

Table 4. Total embodied (direct and indirect) energy use and energy subsidies for agriculture in 2011.

	Rainfed crops	Irrigated crops	Other agricultural activities	Total
Value (MAD millions)				
Energy consumption	2,001	9,047	5,770	16,819
Energy subsidies	− 859	− 4,018	− 2,713	− 7,590
Percentage of value added				
Energy consumption	7.4	26.2	16.0	17.2
Energy subsidies	− 3.2	− 11.6	− 7.5	− 7.8

Total energy consumption and total agricultural subsidies

As shown in Table 4, the total direct and indirect energy consumption for agricultural production amounted to MAD 16.8 billion, the equivalent of 17.2% of the value added by the sector, three times the export revenues from the two main agriculture products exported (tomatoes and citrus; MAD 5.8 billion in 2011) and one and a half times the total agricultural exports (see Figure A3 in the online supplemental data).

Total direct and indirect energy subsidies to agriculture amounted to MAD 7.5 billion in 2011, which is more than the total investment budget of the Department of Agriculture (MAD 6.2 billion in 2011; see Figure A1 in the online supplemental data) and three times direct subsidies for private investment in agriculture (MAD 2.3 billion in 2011).

Energy consumption by the irrigated sector was MAD 9 billion, i.e. 4.5 times that of the rainfed sector, even though the rainfed area is 6 times the irrigated area. Direct and indirect subsidies for energy in the irrigated sector were also 4.5 times those in the rainfed sectors.

Conclusion and perspectives

In Morocco, irrigation has been a major component of the development policy pursued by the government in the last half-century to prevent climatic risks and to ensure minimum food security. Investments in irrigation led to the rapid growth of agriculture and improved the sector's contribution to the national economy. Now that most of the available water resources have been mobilized, the current strategy, particularly through the National Irrigation Water Saving Program Support Project, provides substantial subsidies for the conversion of existing irrigation systems (sprinkler and gravity) to localized irrigation systems, which are assumed to be water-saving techniques. We estimate the subsidies to irrigated crops by the new agricultural strategy at MAD 111,000 per ha, on top of the subsidies allocated to other agricultural investments (machinery, buildings, plantations, greenhouses, etc.). This current policy for reconversion to localized irrigation is an expensive investment but will benefit only the one-third of farmers who already have access to irrigation and water. Moreover, these increased efforts led to intensification and expansion and resulted in an increase in energy consumption in the sector.

The construction of an SAM disaggregated into groups of crops per production system (rainfed and irrigated) and sub-branches of the energy sector, plus the computation of multipliers of production, showed that indirect subsidies, in the form of energy subsidies, exceed direct agricultural investment subsidies. Total direct and indirect consumption of energy in agriculture amounted to MAD 16.8 billion, the equivalent of 17.2% of the value added by the sector and one and a half times the total agricultural exports. Subsidies involved in energy consumption by agriculture amounted to MAD 7.5 billion in 2011, which is more than the total investment budget of the Department of Agriculture. The results also show that multipliers of production, value added and household income are higher in rainfed than in irrigated agriculture. Taking into account the multiplier effects of agriculture, investment in rainfed agriculture would be more profitable for the Moroccan economy. Moreover, irrigated agriculture increases the energy import bill and increases the energy dependency of the country.

During the last half-century, policies in favour of rainfed areas have been disregarded by policy makers although two-thirds of the farmers practise rainfed agriculture. Even if irrigation is an important component of Morocco's agricultural production, the current agricultural policy should be more balanced in favour of rainfed agriculture. As shown here, investment in rainfed agriculture can now be as profitable as investment in irrigated agriculture. What is more, even if irrigation can secure part of the agricultural production, rainfed agriculture has also a high potential to contribute to food security and to poverty alleviation. Productivity could be improved in rainfed areas by better management of climate risk through improved management of soil moisture, zero tillage, drought-resistant varieties, etc. Of course, such improvements also imply investments and funding in long-term research programs.

The new cross-sectoral approach described in this article highlights interdependencies between water, energy and food security. The multipliers of production can measure the direct and indirect effects of production by a given production unit on factor income, institutional income (households and the state), investment and imports, and tax and subsidy. This original approach could help overcome one-sided sectoral thinking and sector-biased planning, management and implementation. The approach calls for greater policy consistency and links different policy levels and sectors to meet the challenges associated with the distribution of scarce resources. Such a tool could help design more efficient policy frameworks. However, to be fully implemented, it requires reliable statistical data, particularly the coordination of data collection, and a better assessment of the links between different production activities and the impacts of different policy alternatives. Finally, this SAM could help decision makers as a first step in the implementation of more coherent water, energy and agricultural policies.

Disclosure statement
No potential conflict of interest was reported by the authors.

Supplemental data
Supplemental data for this article can be accessed at http://dx.doi.org/10.1080/07900627.2015.1036966

References

Bazilian, M., Rogner, H., Howells, M., Hermann, S., Arent, D., Gielen, D., . . . Yumkella, K. K. (2012). Considering the energy, water and food nexus: Towards an integrated modelling approach. *Energy Policy*, *39*(12), 7896–7906.
Bell, C., Hazell, P., & Slade, R. (1982). *Project evaluation in regional perspective* (p. x+326). Baltimore: Johns Hopkins University Press.
Besada, H., & Werner, K. (2015). An assessment of the effects of Africa's water crisis on food security and management. *International Journal of Water Resources Development*, *31*, 120–133. doi:10.1080/07900627.2014.905124

Defourny, J., & Thorbecke, E. (1984). Structural path analysis and multiplier decomposition within a social accounting matrix framework. *The Economic Journal*, *94*, 111–136. doi:10.2307/2232220

Diao, X., Doukkali, R., & Bingxin, Y. (2008). Policy options and their potential effects on Morocco small farmers and the poor facing. *Word Food Prices IFPRI*.

Doukkali, R. (2012). Elaboration d'une méthodologie de construction d'un compte satellite de l'agriculture et d'une matrice de comptabilité sociale pour le secteur agricole. Rapport de la Phase 2 Agenda de Partage pour le Progrès, Fellah Conseil.

Doukkali, R., Flichman, G., & Faurès, J. M. (2004). Impacts de l'agriculture irriguée au Maroc, Administration Générale du Génie Rural. Ministère de l'agriculture, du développement rural et de la pêche maritime. Edité par la Division de la mise en valeur des terres et des eaux (AGL), Organisation des Nations Unies pour l'alimentation et l'agriculture (FAO). Rome 2004 (pp. 52).

FAO. (2012). Social accounting matrix for analyzing agricultural and rural developmentpolicies. Conceptual aspects and examples, 22p, by Lorenzo Giovanni Bellù, Retrieved from http://www.fao.org/docs/up/easypol/936/sam_policy_impact_analysis_130en.pdf

Granit, J., Jägerskog, A., Lindström, A., Björklund, G., Bullock, A., Löfgren, R., & ... Pettigrew, S. (2012). Regional options for addressing the water, energy and food nexus in Central Asia and the Aral Sea Basin. *International Journal of Water Resources Development*, *28*, 419–432. Special Issue: Water and Security in Central Asia: A Rubik's Cube - doi:10.1080/07900627.2012.684307

High-Commission for Planning. (2013). Provisional National Accounts 2012 (basis 1998). (36 pages, in French).

Hoff, H. (2011). Understanding the nexus (Background paper for the Bonn2011 Nexus Conference).

Leontief, W. (1986). *Input-output economics*. New York, NY: Published by Oxford University Press.

Malik, R. P. S. (2002). Water-energy nexus in resource-poor economies: The Indian experience. *International Journal of Water Resources Development*, *18*, 47–58. doi:10.1080/07900620220121648

MAPM (Ministère de l'Agriculture et de la Pêche Maritime). (2012). L'Agriculture marocaine en chiffres, 30 p. Rapport du ministère de l'Agriculture, de la Pêche Maritime. Retrieved from http://www.agriculture.gov.ma/sites/default/files/agriculture-en-chiffres-2012.pdf

MEF (Ministère de l'économie et des finances). (2012). Projet de loi de finances pour l'année budgétaire 2012, Royaume du Maroc. Retrieved from www.finances.gov.ma

Miller, R., & Blair, P. (2009). *Input-output analysis: Foundations and extensions*. New York: Cambridge University Press.

Pyatt, G., & Round, J. (1985). *Social accounting matrices: A basis for planning*. Washington, DC: The World Bank. Retrieved from http://documents.worldbank.org/curated/en/1985/09/439689/social-accounting-matrices-basis-planning

Scott, C. A., & Shah, T. (2004). Groundwater overdraft reduction through agricultural energy policy: Insights from India and Mexico. *International Journal of Water Resources Development*, *20*, 149–164. doi:10.1080/0790062042000206156

Smil, V. (2000). Energy in the twentieth century: Resources, conversions, costs, uses, and consequences. *Annual Review of Energy and the Environment*, *25*, 21–51. doi:10.1146/annurev.energy.25.1.21

System of national Account. (1993). Prepared by the Inter-secretariat working group on national accounts. Retrieved from https://unstats.un.org/unsd/nationalaccount/sna.asp (pp. 838).

Van Vuuren, D. P., Nakicenovic, N., Riahi, K., Brew-Hammond, A., Kammen, D., Modi, V., & ... Smith, K. R. (2012). An energy vision: The transformation towards sustainability—interconnected challenges and solutions. *Current Opinion in Environmental Sustainability*, *4*, 18–34. doi:10.1016/j.cosust.2012.01.004

Waughray, D. (Ed.). (2011). *Water security: The water–food–energy–climate nexus*. Washington, DC: Island Press.

World Bank. (2013). Poverty and social impacts analysis of the Moroccon green growth policy. Energy Axis, a general equilibrium. Département du Développement durable (pp. 41).

WEF (World Economic Forum). (2011). *Global risks 2011* (6th Edition). Geneva: World Economic Forum.

Yang, L., Thurlow, J., & Lahr, M. (2012). The declining role of households in the sustaining China's economy; structural path analysis for 1997–2007. working paper n° 2012/83, UNI WIDER.

Impact of the Syrian conflict on irrigated agriculture in the Orontes Basin

Hadi H. Jaafar[a], Rami Zurayk[b], Caroline King[c], Farah Ahmad[d] and Rami Al-Outa[d]

[a]Department of Agriculture, Faculty of Agricultural and Food Sciences, American University of Beirut, Lebanon; [b]Department of Landscape and Ecosystem Management, Faculty of Agricultural and Food Sciences, American University of Beirut, Lebanon; [c]School of Geography and the Environment, Oxford University Centre for the Environment, UK; [d]Department of Agriculture and Department of Landscape and Ecosystem Management, Faculty of Agricultural and Food Sciences, American University of Beirut, Lebanon

The impact of conflict on irrigated agriculture and consequently summer crop production within conflict-affected agricultural lands was observed in the Orontes Basin. Water and energy use were reconfigured through a transition from rainfed to irrigated agricultural production over the past 20 years, but have been disrupted as the Syrian war has unfolded since 2011. Remotely sensed vegetation indices were used to determine irrigated summer crop yields during the year 2013. Findings suggest that irrigated agricultural production dropped between 15% and 30% in the Syrian portion of the basin in 2000–2013, with hotspots identifiable in Idleb, Homs, Hama, Daraa and Aleppo. The developed approach demonstrated effectiveness in quantifying and geolocating hotspots where conflicts have the strongest impact on agricultural water use, agricultural production, and eventually support relief and regional agricultural reconstruction in this and other conflict regions.

Introduction

Remote sensing can provide a prompt and relatively accurate assessment of agricultural water use and productivity in conflict zones, providing insights on the food security status of the affected communities. Considerable efforts have been spent in developing agricultural and crop yield change methods using remotely sensed data (see e.g. Basso, Cammarano, & Carfagna, 2013). While remote sensing and satellite imagery uses for civil and agricultural applications have recently increased, there are fewer instances in which satellite imagery has been used to study the effects of war on changing human and natural landscapes and food security. Examples include imagery-based crisis identification used by various agencies, including the UN and non-governmental agencies (Marx & Goward, 2013; Marx & Loboda, 2013), for assessment of violation of human rights. Other examples include observation of villages in Sudan destroyed by war (HIU, 2004) and conflict-led rural abandonment of agricultural lands in the two-year war in Kosovo (Terres, Biard, & Darras, 1999), and Bosnia (Witmer, 2008) and Darfur using MODIS (Moderate Resolution Imaging Spectroradiometer) imagery and SPOT (Satellite Pour l'Observation de la Terre) imagery.

This investigation focuses on the Orontes Basin, an international river basin shared by Lebanon (8%), Syria (56%) and Turkey (36%) (Wolf, Natharius, Danielson, Ward, & Pender, 1999). Over the past 20 years, this basin has undergone a transition from rainfed to irrigated crop production. The resulting changes in patterns of water and energy use for food production were particularly rapid in Syria. The percentage of irrigated agriculture in Syria increased from 28.55% to 44% (from 1 million to 1.5 million ha) from 1994 to 2004, although the total farmed area dropped by 13% (from 3.8 million to 3.3 million ha).

The evolving regional dynamic of the water–food–energy nexus changed rapidly with the outbreak of war in Syria in March 2011. By mid-2014, it had led to the displacement of more than six million Syrians. The majority of the displaced population were rural and depended on agriculture for their livelihood. Since the onset of the war, agricultural production in Syria has dropped severely, with millions in rural areas requiring emergency assistance and support. With the possible exception of the coastal areas, most of Syria's rural areas were significantly affected. News reports from the Food and Agriculture Organization of the United Nations have suggest that agricultural production has been halved in some areas (FAO, 2013)

Due to the security problems and the difficulty and the high risk involved when entering the affected areas, little research has been done to quantify the effect of the conflict on agricultural production and water use within Syria. Jaubert, Munger, and Bosch (2014) conducted fieldwork in the Qusair region in Homs and report a decline of production of 70% attributed to conflict. From early 2014, government and rebel forces are each believed to have controlled about 40% of the basin, while the remaining 20% were combat zones (Jaubert et al., 2014). This makes both sides equally capable of implementing a scorched-earth policy by cutting the flow of the irrigation water. It must be noted, however, that just 37% of the irrigated land in Homs and 10% in Hama uses water from government irrigation projects where the flow can be controlled (Table 1); over 60% of the water originates from individual wells which cannot be controlled by the state forces. These data were confirmed by the fieldwork of Jaubert et al. (2014, p. 18) specifically for the Qusair region.

Remote-sensing analysis of satellite imagery and other remotely acquired data offers a safe and promising tool that can aid in observation and quantification of changes in agricultural water management and support preparedness for emergency response activities. Available satellite sensors provide a means for the indirect assessment of agricultural changes via estimates of actual evapotranspiration (ET), Net Difference Vegetation Index (NDVI), Enhanced Vegetation Index (EVI) and other indices. These

Table 1. Irrigated areas (in ha) according to sources of irrigation water in studied regions.

	Irrigated lands (ha)			
	Pressurized irrigation	Irrigation by gravity		
Governorate	Rivers and springs	Government irrigation projects	Wells	Total irrigated land
Homs	9,457	19,902	25,878	53,857
Hama	5,576	6,863	56,871	69,044
Idleb	4,490	7,396	43,489	54,951
Quneitra	590	1,389	2,528	4,331
Aleppo	37,075	60,726	98,175	193,059
Daraa	4,127	18,107	12,813	31,318
Rural Damascus	17,363	4,207	48,477	66,106

sensors collect data at a coarse resolution (250 m to 1 km) and may not accurately reflect the ground situation, especially for small agricultural clusters.

With the presence of high-resolution satellite sensors such as Geo-eye, WorldView 1 and 2, IKONOS and QuickBird, satellite image data available at various spatial, spectral and temporal resolutions allow agriculture and crop assessment. Parameters include crop health, change detection, environmental analysis, irrigated landscape mapping, yield determination and soils analysis. The major constraint regarding analysis using imagery from such sensors is the high cost associated with the imagery (prices can reach $38/km^2 for tasking, with minimum orders of 25 km^2 per area of interest). Another constraint is the huge amount of computer storage and memory required for analysis of these images. The purpose of the research is the development of a GIS-based crop yield model based on vegetation indices derived from freely available remote-sensing data as a support tool for parties engaged in planning for humanitarian response. Such a model would be important not only for use in times of war but also for reducing the impacts of natural disasters such as recurring droughts.

The model provides quantitative data on the production of irrigated summer crops in the Orontes Basin of Syria, one of its main agricultural regions but currently at the centre of ongoing military activities.

This could support preparedness and mitigation interventions to enhance the resilience of rural communities. It will also help delineate zones in need of emergency relief.

Methodology

Study area background and scene selection

In April 2011, Syrian anti-government protests slowly turned into an armed civil conflict between the Syrian army and anti-government forces. By January 2014, the war had caused the death of more than 130,000, the displacement more than 6 million Syrians and severe damage to the economy and infrastructure of Syria. Estimates for damages to the agricultural sector are more than $1.8 billion. Syrian agricultural production is reported to be dropping as conflict continues, with wheat and barley production showing a 55% drop, vegetables 60%, and fruit trees and olive oil production 40% (FAO, 2013). Aside from the widespread lack of security in most governorates, Syrian agriculture in the last three years has suffered from increased fuel costs, destroyed and looted machinery, transportation problems and destruction of irrigation infrastructure. There are also reported cases of decrease in harvested lands. Agricultural trade has also been affected by sanctions.

The studied areas lie in parts of the Orontes River basin in western Syria and in north-eastern Lebanon (Landsat Scene Path 174 Row 036). Three studied regions are near the cities of Homs and Qusair (home to major field battles in May 2013); one is north of Hama (home to a major conflict in spring of 2012).

Additional regions in Lebanon were selected to disentangle the effect of conflict from that of other agro-ecological factors. The main comparison unit is in the Qaa area of Lebanon, adjacent to the Qusair region in the Homs District and part of the Orontes Basin. The Mid-North Beqaa area, falling within the Litani River basin and unaffected by the war in Syria, was added as a 'placebo' to address the possible interdependency of the Qaa and Qusair sites, according to the procedure for comparative analysis of small-n samples developed by Glynn and Ichino (2014). This choice of method is supported because agricultural integration between the two areas is minimal. Production in the two regions aims at different markets, one essentially Syrian and the other Lebanese. Some minimal overlap could have taken place before 2011 as inputs were smuggled from Syria to Lebanon (essentially subsidized diesel). The only possible effect is the overflow of labour from Syria to El Qaa, but this is a common phenomenon, dating back to the mid-twentieth century, as Lebanon has always been the main destination for Syrian seasonal migrant labour and labour shortages were never experienced.

The long-term production records for El Qaa are stable, indicating no effect of the war in Syria on agricultural production. The long-term records for adjacent Qusair show a dip during the period of study. The only difference is the active conflict in Qusair. According to John Stuart Mill's method of difference (quoted in Glynn & Ichino, 2014, p. 5):

> If an instance in which the phenomenon under investigation occurs, and an instance in which it does not occur, have every circumstance save one in common, that one occurring only in the former; the circumstance in which alone the two instances differ, is the effect, or cause, or a necessary part of the cause, of the phenomenon.

If one agrees with the principle that a regional state of war is different from an active conflict, and that the two locations are otherwise similar (in agro-climatic conditions), then this directly implies that the conflict is the cause of the observed difference in crop yields between El Qaa and Qusair.

Figure 1 shows the studied areas for the assessed Landsat scenes. The study area for the MODIS NDVI, EVI and Drought Severity Index (DVI) data is the irrigated zones within the country of Syria. All the study areas are distinctively agricultural. All areas rely on irrigation for agricultural production, because the summer growing period (June–August) is characterized by zero effective rainfall. Evaporation in April and May is high enough to deplete the winter residual of the available soil moisture within the root zone of summer crops. The Orontes River and associated springs and irrigation projects are the major source of water for these areas. The areas are analyzed for recent marked changes in atmospherically corrected reflectance from the red and near-infrared bands of the acquired Landsat imagery.

The locations of the study areas were selected based on the following rationale:

(1) The areas are heavily farmed, cropped and irrigated during normal years from storage reservoirs and from the Orontes River. Field battles amongst the conflicting forces had occurred in the Syrian territories and are reported to have disrupted farming activities in those areas. The selected Lebanese areas were not directly affected by major Syrian conflict during the analyzed period (2000–2013), allowing us to consider these areas as comparative units (with the exception of the Qaa region).

(2) The area on the Lebanese–Syrian border (Qaa) is contiguous with the Qusair Area in the Syrian territories, the examination of which will determine whether there was an impact of the Syrian crisis on agricultural activities in that zone.

(3) The study areas lie in the same climatic zone (arid to semi-arid), with similar topographic, vegetative and crop-type characteristics.

(4) The 'placebo' area was delineated based on the existence of agricultural lands within the same Landsat scene, to preserve the temporal resolution of the analysis.

(5) The climate of the region is arid to semi-arid. Annual average precipitation ranges from 200 to 600 mm/y. The wet season starts in October and ends in April. During the dry season, the eastern part of the selected Landsat scene (inland) is mostly cloud-free. The western part (mostly mountains and coastal area) sometimes has significant cloud cover.

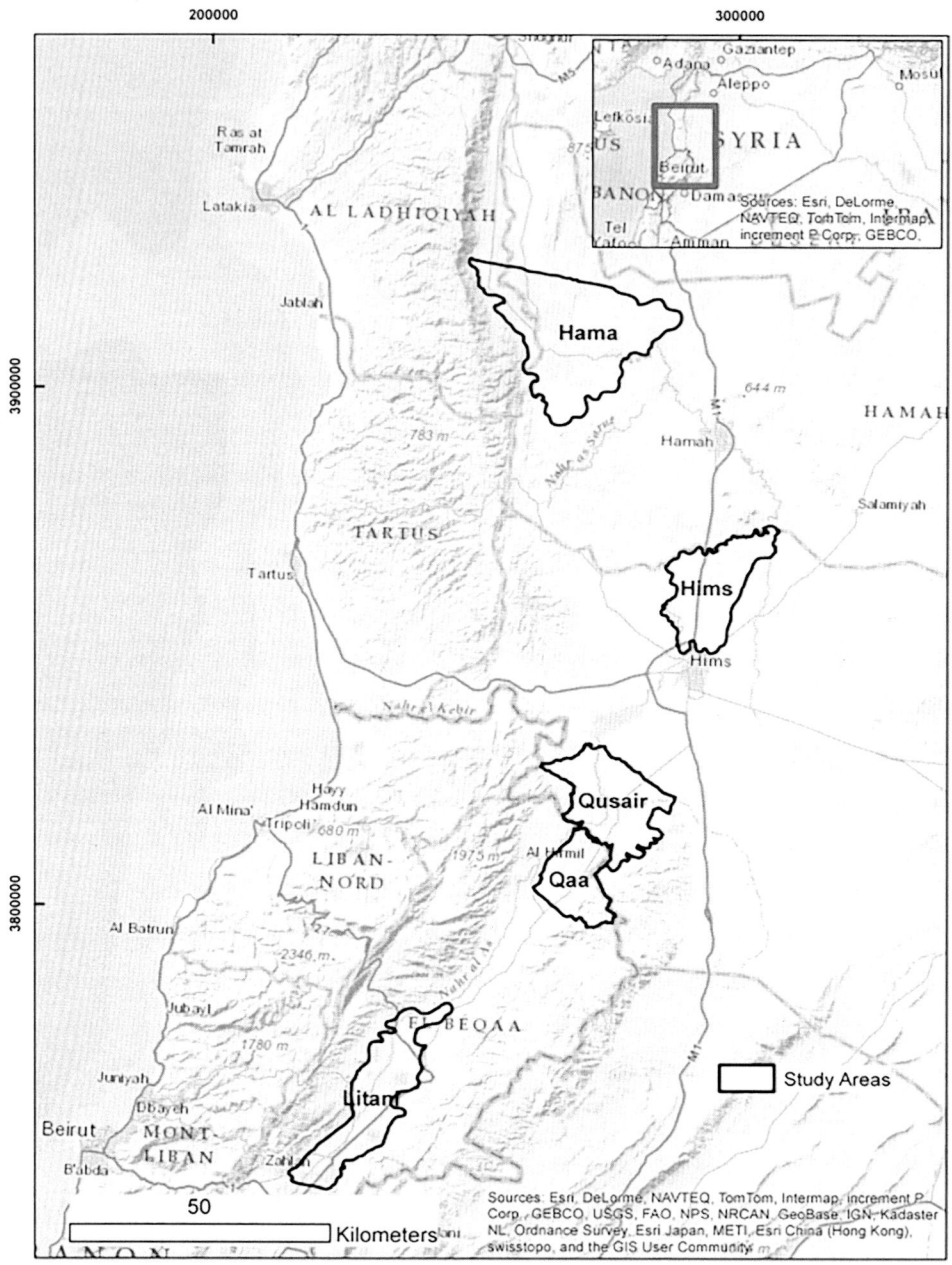

Figure 1.   Pilot study areas for the fraction of vegetation analysis.

Landsat 5 Thematic Mapper (TM), Landsat 7 Enhanced Thematic Mapper Plus (ETM +) and Landsat 8 Operational Land Imager (OLI) scenes of the last 10 years were analyzed for fraction of vegetation cover (FOV) as derived from the NDVI. The input data from the developed model include the red and near-infrared bands of the Landsat scene.

Scenes and data processing

A model using remote sensing and GIS was developed to assess the impact of the conflict on irrigated agricultural production using zonal statistics of NDVI, and the land cover change from freely available Landsat imagery (30 m resolution, see Table A1 in the online supplemental data at http://dx.doi.org/10.1080/07900627.2015.1023892).

Satellite-derived ET (Mu, Zhao, & Running, 2011) and the MODIS EVI were also used to quantify effects on agricultural water use and summer agricultural production (after Doraiswamy et al., 2005). Statistical analysis was conducted to detect the significance of changes in these metrics. The model was validated by comparing its output to production data from the pre-war years. The DSI (Mu, Zhao, Kimball, McDowell, & Running, 2013) was used to separate the effect of drought from the impact of conflict through comparison with areas in the same climatic zone in neighbouring Lebanon.

Landsat images from various sensors were collected for the summer season of years 2000–2011 and 2013. A total of 41 Landsat scenes were analyzed (May 2000 to 2013). All of the satellite images were selected to have less than 5% cloud cover and image quality better than 9 for the five study areas. When clouds covered one study area but not the other, the Landsat scene was excluded from the analysis for that area. The images detecting the change in agricultural activities are the L8 images of 2013. Depending on the Landsat mission (Landsat 5 TM, Landsat 7 ETM + or Landsat 8), three different GIS data models are developed. Landsat scenes and other remotely sensed data were downloaded from the US Geological Survey website (http://earthexplorer.usgs.gov/).

The NDVI is derived from Landsat 5 and 7 bands by converting calibrated 'digital numbers' of the scenes to absolute units of at-sensor spectral radiance, to top-of-atmosphere reflectance and then to NDVI using the metadata file and the necessary computational algorithms. To ensure the same scaling for the studied scenes, raw digital numbers of the bands are converted to at-sensor spectral radiance based on the procedure described in Chander, Markham, and Helder (2009). FOV was derived from the NDVI estimates (see Carlson, Gillies, & Schmugge, 1995; Gillies, Kustas, & Humes 1997). Zonal statistics for an area of interest, with means and standard deviations, were calculated in ArcGIS Spatial Analyst.

For Landsat 8, reflectance for Bands 4 and 5 were first calculated using the reflectance scaling coefficients provided in the product metadata file. The FOV is derived from the scaled NDVI between bare soil and full vegetation (Gillies et al., 1997). This scaling overcomes some of the limitations incurred by comparing the NDVI of different images, like atmospheric interference and variations in soil brightness and background scattering. Landsat 8 launched May 2013; a total of six Landsat scenes were analyzed (late May to August).

The derived FOV was used to calculate actual ET in 2013. FOV in 2013 was compared to FOVs in 2000–2011 to determine any statistical significance of differences. Zonal statistics for the other parameters (DSI, MODIS NDVI and EVI) were calculated. Trends in seasonal ET (2000–2012) are also presented to separate conflict effects from drought. DSI data (5 km) were analyzed for the period 2000–2011, MODIS NDVI and EVI data for the period 2000–2013. GIS processing and spatial zonal statistics were calculated, followed by statistical analysis for significance.

The main index that is indirectly derived from the red and near-infrared bands is the FOV. Although the derived relationship between the FOV and NDVI does not have a purely physical basis, it has been verified and documented in several works (Gillies et al., 1997). The stability of the NDVI index depends on several factors. The NDVI index indicates plant vigour. It usually exhibits strong seasonal dependence, mainly in late winter and early spring. In agricultural areas, NDVI sharply drops in summer. A high NDVI is related to healthy vegetation and strong plant vigour, while a low NDVI indicates stressed vegetation and low plant vigour, mainly low ET ('cold pixels').

To minimize the seasonal effect of precipitation, the main months that are compared are June, July and August. To avoid signal instability, NDVI is computed and compared from scenes taken within the same months for the period of record. The time difference between the images compared is a multiple of one year. Intra-annual local weather variability and changes in air masses can cause differences in remotely sensed NDVI that cannot be purely attributed to conflict effects. The compared images are mainly in the dry season (when irrigation is indispensable for crops). Average NDVI is computed for the study areas for the months of June, July and August. Six Landsat 8 images are used and compared to the averages of the zonal statistics of the reference years within the period of interest. In previous Landsat missions, the metadata file has parameters for the radiometric scaling necessary to convert digital numbers to radiance and then to reflectance. The bandwidth of the TM and ETM infrared band is 0.76–0.90, while that of the Landsat 8 OLI is narrower (0.85–0.88 nm). NDVI and FOV observations for agricultural fields are statistically tested using Student's t-test, two-tailed, to determine whether the observation is significantly different from the reference years studied.

The method of comparing FOV observation to the mean of its previous observations increases the sensitivity of the algorithm by eliminating differences between agricultural fields such as local land cover and the density of urban settings, which change little by little with time.

Drought severity index and ET

MODIS-derived ET was derived using zonal statistics for the study areas and means where calculated for all the months from 2000 to 2012. The derived ET is the global terrestrial evapotranspiration that is determined from Global World Meteorological Observations and MODIS NDVI (Mu et al., 2011). Monthly means are derived the summer months of June, July and August, and trends are analyzed for significance for the period of study. MODIS-derived ET is used to calculate a new DSI. The DSI is calculated by summing standardized ET cell by cell with standardized NDVI, and then standardizing the sum. For a detailed analysis of the DSI algorithm see Mu et al. (2013).

Enhanced Vegetation Index analysis

NDVI might be subject to errors from soil reflectance and lack of atmospheric normalization. The EVI is a modified form of NDVI that has a soil adjustment factor and some correction for the red band due to aerosol scattering in the atmosphere. EVI is used to detect changes in irrigated lands in the major agriculture-producing governorates. EVI metrics were analyzed by the following procedure. Monthly 1 km gridded MODIS EVI data were downloaded for all of Syria and Lebanon for the summer months of June–August for the years 2000–2013. Standardization was performed on a cell-by-cell basis according to the following. Let v be the value of the parameter of every cell i in a raster for a time period t. For all i,

$$Z(i) = \frac{\{v_{it} - \mu(v_{it})\}}{\sigma(v_{it})}$$

where $\mu(v_{it})$ is the temporal mean for the period 2000–2013 for every cell in the raster for every cell i in time period t; σ is the temporal standard deviation for every cell in the raster for every cell i in time period t; and Z is the resulting dimensional index, in which trends and seasonality have been removed from the series. Zonal statistics were conducted on the metrics above to calculate the mean SEVI for the political units (governorates) of the study area. The means of EVI for conflict years 2012 and 2013 were statistically compared to the mean of 2000–2011 in each governorate.

Time series spatial means for irrigated zones for political units where calculated using GIS and regressed against time series of government-reported cropping data within the governorates for the last decade.

Results

NDVI and FOV analysis for pilot areas

Landsat-derived mean fractional vegetation cover values (FOV) for the study area in June, July and August (JJA) of 2013 were compared to the mean values for 2000–2011. A significant reduction was found in FOV as compared to the baseline mean in the Qusair area, reflecting a proportional reduction in consumptive water use. This is an indication of a net reduction in the irrigated area (-15%) and/or an increase in crop water stress. This decline was not registered for the adjacent Lebanese Qaa area or for the Litani 'placebo' zone, indicating that the most probable cause for the downtrend is the violent conflict that shook the region during the summer of 2013. These results are in agreement with those of Jaubert et al. (2014, p. 18), who report a decline in yields of 60% in Qusair during the same period. These findings are further confirmed by the mean values of MODIS-derived NDVI readings for the same dates (Table A2 in the online supplemental data), which show a decline of 15% for the Qusair region in the summer of 2013. During the same period, neither of the Lebanese regions, Qaa or Litani, showed any significant reduction. In fact, those areas exhibit an increase in FOV.

In northern Homs, FOV for June and July were significantly lower at a 90% confidence level, but not at 95%. Overall, in the summer months of 2013, FOV in the northern Homs part of the Orontes Basin was statistically equal to previous years. In Hama, 2013 values were significantly lower for the month of June, but not for the later summer months. Over the summer of 2013, mean FOV for Hama was statistically equal to previous years.

ET and drought severity for Syria and Lebanon

The three-year moving average of summer ET anomalies for the five study areas is presented in Figure A1 (in the online supplemental data). The period 2000–2008 shows a decreasing ET trend followed by an increase in mean spatial ET anomalies after 2008. The MODIS ET has been validated via flux-tower data on several sites worldwide, and it is believed to be accurate within 25–30%. The wider range of summer anomalies calculated for the Hama region is believed to be due to the large irrigation activities and heavy summer agriculture in that zone. A decline in these activities will produce a sharp decrease in actual evapotranspiration, and a sharp decrease in actual evapotranspiration can be attributed to a decrease in irrigated areas, as compared to other zones with a lower average ET. This will cause a sharper anomaly. Spatial means for DSI are shown in Figure 2. Higher DSI indicates a 'wetter' summer. The DSI trend is very similar to the MODIS ET trend (DSI was derived from MODIS ET/potential ET and 1 km MODIS NDVI). An increased mean summer DSI is synchronous with an increased summer ET in

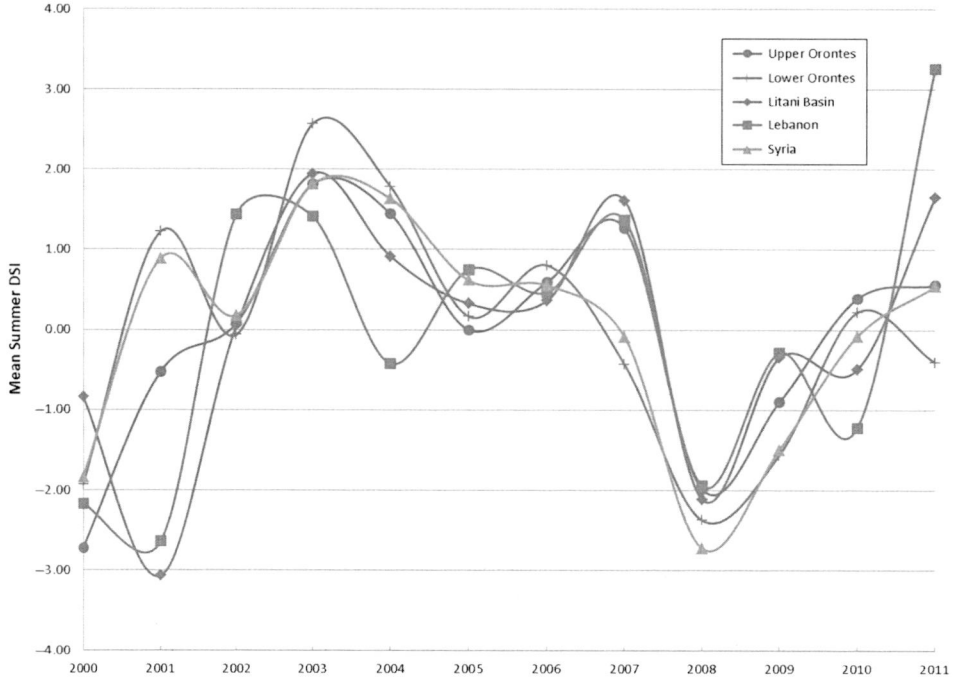

Figure 2. Spatial mean of summer Drought Severity Index for selected areas.

2008–2011. No drought trends are significant within Lebanon, Litani Basin, or Upper Orontes. The spatial DSI analysis for the 12 years for Syria and Lebanon is shown in Figure A2 (in the online supplemental data). The year 2004 is drier than average across the Eastern Mediterranean coast, while 2008, 2009 and 2010 (pre-conflict years) are drier than average within the inner arid regions. Because this drought index relies on NDVI and ET calculations, it is dependent on mainly two factors: weather and vegetation. Vegetation is affected by rainfall and agricultural activities, and the relationship between NDVI and actual ET is predictable in most cases: an increase in NDVI implies healthy vegetation, a higher leaf-area index and higher absorption of photosynthetic active radiation, causing ET to be higher for constant weather (as more green water is available for transpiration). When weather parameter values increase, ET will increase for a constant NDVI, eventually causing NDVI to decrease later in the season if water supplies are limited, hence limiting ET values afterwards.

EVI analysis

EVI was calculated for the five pilot areas as well as for the summer-irrigated lands and the administrative units of Syria. Recently MODIS EVI has been used to predict crop yields, with good results (Atzberger, 2013; Potgieter, Apan, Dunn, & Hammer, 2007). The major agriculture-producing zones are the Al-Ghab Valley and the Orontes Plain. Figure A3 (in the online supplemental data) shows spatial EVI anomalies for 2012 and 2013 as compared to the mean of 2000–2011. The regions suffering from the highest EVI drop are northern Lathikiya (on the Syrian–Turkish border), the banks of the Orontes River, parts of Idleb, and Aleppo. It is noted that the same regions are affected, but the wet year of 2013 had a much lower SEVI, especially within the agricultural areas intersecting with combat zones. In 2013, areas of Damascus (Ghouta region, location of major combat in 2013), the Orontes Basin, the Turkish border at Lathikiya (another major combat zone), and the borders of Idleb and Hama are severely affected. The same areas appear to be affected when analyzing SEVI time series for the political units of Syria. Changes in agricultural lands in Quneitra (Occupied Golan Heights) and Sweida (little or no conflict) are barely noticed.

It is evident that during the years 2012–2013 there is a sharp drop in SEVI. Also, it is much more pronounced in 2013 than in 2012; this is due to the intensification of the conflict in 2013. Within the arid zones of Syria, the main vegetation cover is solely irrigated agriculture, as there is too little rainfall to sustain summer vegetation. The SEVI is particularly helpful in identifying agricultural areas, as it is these areas that suffer from stress in a variable-climate situation. To determine the relationship between EVI and agricultural production within the studied irrigated areas, summer crop yield data for the years 2000–2011 were plotted against the spatial mean of the sum of summer EVI (June, July and August) for the pre-conflict period. Regression coefficients for the relationship were determined. F-statistics were calculated to determine the significance of the trend at the 5% level.

Figure 3 shows the derived relationships between summer crop production and remotely sensed summed summer EVI for the major agricultural governorates in Syria.

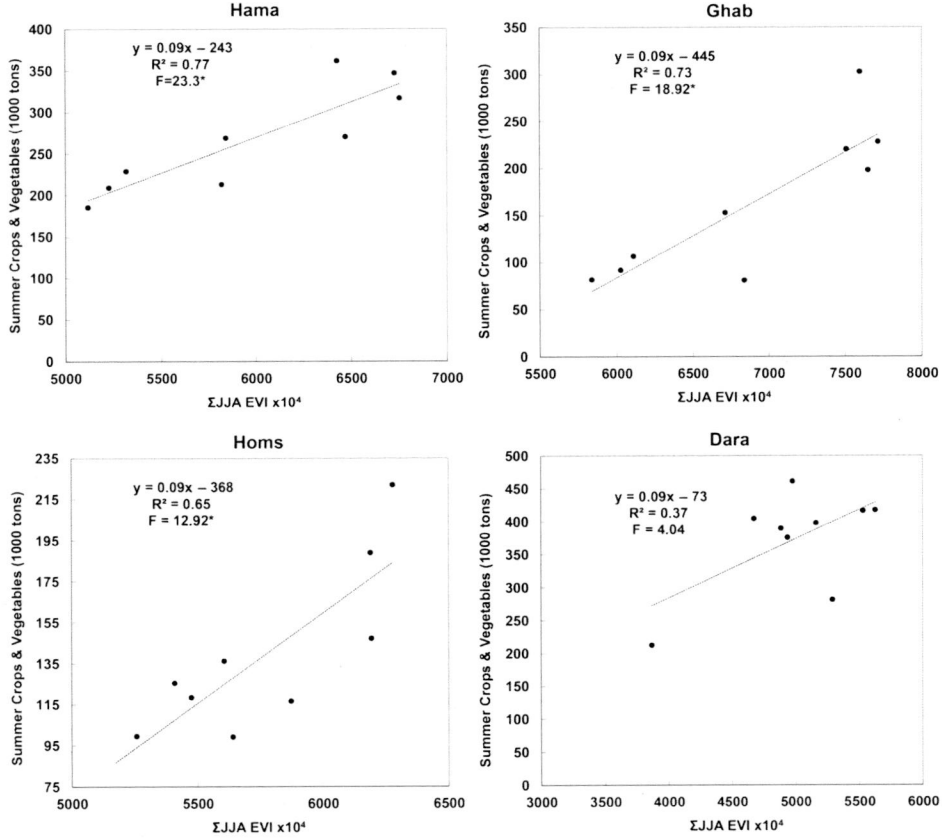

Figure 3.   Relationship of summer agricultural production to cumulative summer EVI.

A significant positive linear relationship is seen in three cases (irrigated zones of Homs, Al-Ghab and Hama). The highest F-statistic is for the Hama irrigated area and Al-Ghab Valley, both lying within the Orontes River basin and both characterized by high summer production. Sum of EVI predicts production in these regions with relatively high accuracy. These relationships can be used to estimate the changes in summer production for 2012 and 2013, where crop data are not reliable (2012) or are lacking (2013). EVI mostly dropped in 2013 and 2012, as shown in Table 2. Idleb is affected most, followed by Homs, Hama, Daraa and Aleppo. Although in Idleb, Aleppo and Daraa there is a drop in EVI, there is no significant relationship between EVI and crop production in these areas. The drop in EVI more than doubled in 2013 as compared to 2012. Quantification of this drop will lead to identifying the governorates and agricultural areas that were the most affected by the conflict.

Discussion

The investigation has achieved its aim of developing a simple, low-cost remote-sensing method for evaluating changes in the productivity of irrigated areas in specific zones without the need for 'boots on the ground'. The methodology for remotely assessing the quantitative variation in crop yields at the level of regions and basins is useful, given that the process of statistical data collection of the Syrian state, once quite solid, has been seriously hindered by the conflict and data are therefore either nonexistent or unreliable. The method was effective in identifying areas where agriculture has been adversely affected by the existing conflict.

The findings of the analysis of remotely sensed vegetation indices could provide guidance to humanitarian agencies on where emergency food aid should be focused. It will also be valuable in helping understand the relationships between decreased production, displaced rural population, and possibly shifts of labour to other sectors.

Various remotely sensed parameters related to agricultural water use and agricultural production in Syria's Orontes Basin were analyzed to evaluate the state of agricultural production as the conflict continues to unfold. The percentages of land area under irrigated vegetation cover identifiable from Landsat imagery were found to be significantly lower in 2013 than in previous years.

The indices used show a decrease in agricultural vegetation and eventually lower consumptive water use during the last decade over some parts of the Levant region. Mapping of standardized EVI demonstrate changes in vegetation within summer-irrigated lands. The spatial mean of the summer sum of EVI is highly correlated with reported yields in non-river-irrigation-dependent areas.

Conflict rather than drought is considered to have caused the reduction in agricultural productivity. The year 2013 was marked by intense battles in the area where the changes were detected. On the other hand, although a trend of increasing drought was detected for 2000–2009, this was followed by a trend of decreasing drought (2010–2012), which matches well with climatological observations. There was no significant overall drought trend.

The decline in production is not proportionally related to the decline of available water supply and/or energy for that supply. Rather, these are viewed as two of the factors affecting production. The major factor is the conflict itself, and the possibility of farmers temporarily abandoning their agricultural activities and lands for safety reasons. Identification of the specific causal mechanism(s) for the observed reduction probably cannot be done using remote-sensing methods alone, and is certainly beyond the scope of

Table 2. Changes in EVI and calculated summer crop production for the years 2012 and 2013.

Irrigated lands	2001–2011 mean summer sum of EVI	2012 mean summer sum of EVI	2013 mean summer sum of EVI	2012 % EVI difference from mean	2013 % EVI difference from mean	Reported mean summer production, 2000–2011	Predicted summer production, 2012	2012 predicted summer production minus % difference from reported mean (2000–2011)	Predicted summer production, 2013	2013 predicted summer production minus % difference from reported mean (2000–2011)
Hama	0.60	0.53	0.48	−11.8	−19.5	152	93	−39.0	60	−60.5
Ghab	0.69	0.60	0.54	−13.6	−21.2	148	61	−58.7	25	−83.3
Homs	0.58	0.52	0.45	−11.4	−22.4	141	97	−30.7	39	−72.0
Idleb	0.55	0.49	0.40	−10.1	−26.1	265	303	14.4	276	4.2*
Sweida	0.31	0.31	0.29	0.2	−6.5	51	50	−1.9	42	−17.2*
Aleppo	0.62	0.62	0.50	0.9	−19.5	483	629	30.2	662	37.1*
Daraa	0.50	0.48	0.42	−3.8	−15.5	342	182	−46.8	159	−53.6*

* Not significant at $p < .05$

this work. For insights into the mechanisms underlying the decline in crop productivity in the study area during the period of study, the reader is referred to the work of Jaubert et al. (2014) and Zurayk (2014).

Conclusion

The article has approached the issue of irrigation and agricultural production in war-torn Syria in a new way. Remotely sensed vegetation indices over the Syrian portion of the Orontes River Basin were derived and analyzed. The research findings indicate that irrigated agricultural production dropped between 15% and 30% in the Syrian portion of the basin 2000–2013, with hotspots identifiable in Idleb, Homs, Hama, Daraa, and Aleppo regions of Syria. The remote sensing approach proved effective in identifying and locating losses in agricultural productivity due to conflict, rather than drought. However, the remote sensing tools could not be used to determine the precise causal mechanisms and dynamics taking place between the conflict and other factors, including those related to water and energy management at the local scale. Nonetheless, the effectiveness and the potential for remote sensing to be used to support relief and regional agricultural reconstruction in this and other conflict regions was demonstrated by correlating the analyzed indices to irrigated agricultural production.

Disclosure statement

No potential conflict of interest was reported by the authors.

Funding

This research was supported by funds from the American University of Beirut Research Board and the United States Agency for International Development under the Middle East Water and Livelihoods Initiative, managed by the International Centre for Agricultural Research in Dry Areas. The views expressed in this publication are those of the authors and do not necessarily reflect the views of funding agencies.

Supplemental data

Supplemental data for this article can be accessed at http://dx.doi.org/10.1080/07900627.2015.1023892

References

Atzberger, C. (2013). Advances in remote sensing of agriculture: Context description, existing operational monitoring systems and major information needs. *Remote Sensing, 5*, 949–981. doi:10.3390/rs5020949

Basso, B., Cammarano, D., & Carfagna, E. (2013). Review of crop yield forecasting methods and early warning systems. In *Proceedings of the first meeting of the scientific advisory committee of the global strategy to improve agricultural and rural statistics.* Rome: FAO Headquarters. 18–19 July.

Carlson, T. N., Gillies, R. R., & Schmugge, T. J. (1995). An interpretation of methodologies for indirect measurement of soil water content. *Agricultural and Forest Meteorology, 77*, 191–205. doi:10.1016/0168-1923(95)02261-U

Chander, G., Markham, B. L., & Helder, D. L. (2009). Summary of current radiometric calibration coefficients for Landsat MSS, TM, ETM + , and EO-1 ALI sensors. *Remote Sensing of Environment, 113*, 893–903. doi:10.1016/j.rse.2009.01.007

Doraiswamy, P. C., Sinclair, T. R., Hollinger, S., Akhmedov, B., Stern, A., & Prueger, J. (2005). Application of MODIS derived parameters for regional crop yield assessment. *Remote Sensing of Environment, 97*, 192–202. doi:10.1016/j.rse.2005.03.015

FAO. (2013, January 13). News article. Retrieved from Food and Agriculture Organization of the United Nations website http://www.fao.org/news/story/en/item/168676/icode/

Gillies, R. R., Kustas, W. P., Humes, K. S., et al. (1997). A verification of the 'triangle' method for obtaining surface soil water content and energy fluxes from remote measurements of the Normalized Difference Vegetation Index (NDVI) and surface e. *International Journal of Remote Sensing, 18*, 3145–3166. doi:10.1080/014311697217026

Glynn, A., & Ichino, N. (2014). Increasing inferential leverage in the comparative method: Placebo tests in small-*n* research. *Sociological Methods & Research.* doi:0049124114528879

HIU. (2004). *Sudan (Darfur)–Chad border region confirmed damaged and destroyed villages.* Washington, DC: Department of State, Humanitarian Information Unit. (https://hiu.state.gov/Pages/Africa.aspx).

Jaubert, R., Munger, F., & Bosch, C. (2014). *Syria: The impact of the conflict on population displacement, water and agriculture in the Orontes River basin.* Geneva: Global Program, Water Initiatives Swiss Agency for Development and Cooperation.

Marx, A., & Goward, S. (2013). Remote sensing in human rights and international humanitarian law monitoring: concepts and methods. *Geographical Review, 103*, 100–111. doi:10.1111/j.1931-0846.2013.00188.x

Marx, A., & Loboda, T. (2013). Landsat-based early warning system to detect the destruction of villages in Darfur, Sudan. *Remote Sensing of Environment, 136*, 126–134. doi:10.1016/j.rse.2013.05.006

Mu, Q., Zhao, M., Kimball, J. S., McDowell, N. G., & Running, S. W. (2013). A remotely sensed global terrestrial drought severity index. *Bulletin of the American Meteorological Society, 94*, 83–98. doi:doi:10.1175/BAMS-D-11-00213.1

Mu, Q., Zhao, M., & Running, S. W. (2011). Improvements to a MODIS global terrestrial evapotranspiration algorithm. *Remote Sensing of Environment, 115*, 1781–1800. doi:10.1016/j.rse.2011.02.019

Potgieter, A. B., Apan, A., Dunn, P., & Hammer, G. (2007). Estimating crop area using seasonal time series of Enhanced Vegetation Index from MODIS satellite imagery. *Australian Journal of Agricultural Research, 58*, 316–325. doi:10.1071/AR06279

Terres, J., Biard, F., & Darras, G. (1999). *Kosovo: Assessment of changes of agricultural land use areas for the 1999 crop campaign using satellite data.* Ispra: Space Applications Institute Report, Joint Research Centre.

Witmer, F. D. W. (2008). Detecting war-induced abandoned agricultural land in northeast Bosnia using multispectral, multitemporal Landsat TM imagery. *International Journal of Remote Sensing, 29*, 3805–3831. doi:10.1080/01431160801891879

Wolf, A. T., Natharius, J. A., Danielson, J. J., Ward, B. S., & Pender, J. K. (1999). International river basins of the world. *International Journal of Water Resources Development, 15*, 387–427. doi:doi:10.1080/07900629948682

Zurayk, R. (2014). The fatal synergy of war and drought in the eastern Mediterranean. *Journal of Agriculture, Food Systems, and Community Development, 4*, 9–13. http://dx.doi.org/10.5304/jafscd.2014.042.013

Climate change and food-water supply from Africa's drylands: local impacts and teleconnections through global commodity flows

Mark Mulligan

Department of Geography, King's College London, UK

This article uses the WaterWorld Policy Support System, coupled with a global database for commodity flows, to examine the impacts of AR4 SRES climate change scenarios on Africa's drylands and the commodity flows that originate from them. It shows that changes to precipitation and, to a lesser extent, temperature in Africa's drylands can significantly affect the potential to supply water-for-food locally and internationally. By comparing the geographical distribution of climate change with the supply chain–connected distribution of climate change, it shows how food-water impacts of climate change may affect local dryland populations but also those dependent on these flows from afar.

Introduction

We all know that the climate disruption expected to result from increasing concentrations of atmospheric greenhouse gases will have impacts on the climates of many countries. In all countries, temperatures will increase. In some countries there will be more rainfall than at present, and in others less. Such changes may have profound impacts on the food-water-energy conditions in some of these countries. This will be particularly true for populous dryland countries that already have marginal climates and water-stressed agricultural systems.

Many such countries already mitigate against the aridity or inter-annual variability of their climates by importing significant proportions of their food (and thus food-water) from overseas. Some will have greater food-water security than now, and some will have less than now, but all will have different from now, as local productivity and global markets react to these changes. But the future for food-insecure countries will also depend on the impacts of climate change on food-water availability in the countries they import food from and/or where they have agricultural foreign direct investments. Thus, the food-security future for all countries depends on the impacts of climate change on food-water, locally and also at the supply end of the supply chains that these countries depend on.

Objectives

This article uses the WaterWorld Policy Support System (http://www.policysupport.org/waterworld), coupled with a global database for commodity flows, to examine the impacts of AR4 SRES (Special Report on Emissions Scenarios) climate change scenarios on Africa's drylands and their commodity export flows. We hypothesize that climate change in Africa's drylands will result in changes in the water availability that sustains commodity flows to many countries and that these impacts will differ between commodities and countries affected according to the geographical distribution of production areas for those commodities in relation to the geographical distribution of projected climate change. This is thus the first study to map Africa's climatic future and its impact on water-food and energy both locally and along international commodity supply chains.

WaterWorld is a sophisticated water balance model capable of modelling the impacts of climate change scenarios on the baseline water balance for any region of the world. Full details of the model specification and input datasets are given in Mulligan (2013). WaterWorld is used to compare baseline (WorldClim 1950–2000, Hijmans, Cameron, Parra, Jones, & Jarvis, 2005) climate conditions with an SRES A2a climate scenario, downscaled to 1 km resolution using the delta method relative to the WorldClim climate baseline (see Mulligan, Fisher, et al., 2011). The SRES A2A scenario represents high growth and a global 3.5°C warming relative to 1990 by 2100 (Nakicenovic, 2000). Data from this scenario were used as monthly downscaled general circulation model (GCM) output (Ramirez & Jarvis, 2010) for temperature and precipitation. Only monthly temperature and precipitation changes were examined since most GCMs provide only these variables. Population, land use and international commodity flows are assumed to remain near their baseline levels, since it is rainfall change that we are interested in. For Africa at 1 km resolution this means doing calculations over almost 78 million pixels for each map. Though challenging, this level of spatial detail is important given the complex and detailed geographical patterns and overlap of croplands, pastures and dryland climates and the importance of climate for agriculture. Though the study area is the whole continent of Africa and outlying islands, we focus on Africa's drylands for many of the analyses.

The African distribution of drylands revisited

Drylands are commonly classified on the basis of rainfall or the ratio of rainfall to potential evapotranspiration: P/PET. We first compare these standard metrics for Africa with some novel ones that offer alternative assessments of the distribution of drylands. We focus on annual total metrics rather than those that examine seasonality since seasonal data for some of these variables is less available. Our interest is in the driest lands, so we include the "hyper-arid" and "arid" classifications according to the UNEP classification (United Nations Environment Programme, 1997) and thus assume that arid conditions prevail at less than 300 mm/y of rainfall according to WorldClim rainfall data (Hijmans et al., 2005). These conditions produce the distribution of African arid lands shown in Figure A1(a), amounting to about 15 million km^2. (All figures are in the online supplemental data at http://dx.doi.org/10.1080/07900627.2015.1043046.) Figure A1(b) shows the distribution if we use the UNEP (1997) definition of aridity as P/PET < 0.2 (using the WorldClim-based aridity index of Trabucco & Zomer, 2009) and produces 16 million km^2, with greater extents in south-west Africa and the Horn of Africa.

The former are commonly used measures of climatic aridity. A climatic measure that has not seen use to date for this purpose is cloud frequency. Cloud frequency is likely to be related to aridity since a low cloud frequency means lower rainfall but also higher direct solar radiation loads. Figure A1(d) shows cloud frequency from the MODIS sensor on NASA's TERRA and AQUA satellites for all observations from 2000–2006. The MOD35 cloud product is used to assess the number of cloud observations per 1 km pixel, and these are presented here as a fraction of the total number of observations using the global climatology produced by Mulligan (2006). Examining areas that are cloudy for less than

30% of observations, we find a similar distribution as for the standard aridity index for North Africa, Saharan and Sub-Saharan Africa, but a significant extension in southern Africa, south-west Madagascar, Kenya and Tanzania. The overall extension increases to 17 million km^2.

Areas classified as arid are characterized by a low biological potential of the land, but where there are flows of water from upstream, extensive irrigation or compensating environmental conditions such as high inputs of fog, this reduction in productivity may not be observed on the ground. We can thus also look to estimates of vegetative productivity itself to characterize true drylands. Mulligan (2009) produced a global data-set for mean dry matter productivity. This is calculated from analyses of dry matter production data made available by VITO (http://www.geoland2.eu/) every 10 days over the 11-year period 1998–2008 from 1 km resolution SPOT-VGT NDVI data. Dry matter productivity represents the daily growth of standing biomass. Figure A4(c) shows productivity less than 2000 Dg/ha per day (0.6 kg/m^2 per year). These levels of productivity clearly have a very similar distribution to the standard aridity index (also covering about 17 million km^2). But this aridity index extends into the Great Lakes since the SPOT-VGT data consider only terrestrial plant productivity and lake values are thus low. Bright spots also include the Nile River. Though these areas and their populations would be considered arid in the climatically-based characterizations, they are not so on the ground because of the presence of rivers. This article will continue using the standard aridity metric (see Figure A2), but readers need to bear in mind that there is no single best definition and that climatically defined drylands may not all be hydrologically or productively dry even if they are climatically dry.

Aridity in Africa and the water–food–energy nexus

Within these drylands we now examine the distribution of land used for food (i.e. croplands and pastures), water and energy (as represented by the watersheds of dams) and the size of the local populations to be supplied with food, water and energy. The croplands and pastures map of Ramankutty et al. (2008) was used to represent current croplands and pastures for all croplands (fractional cover) and pastures (fractional cover), all of which were masked by the aridity metric to represent drylands only. Ramankutty et al.'s data combine agricultural inventory data and satellite-derived land cover data to map land use to form the only comprehensive characterization of the distribution of global agriculture as of the year 2000. Figure A3(a) shows croplands in arid zones of Africa, indicating a coverage of 2.7 million km^2 (8.9% of Africa), home to 110 million people (13% of Africa), calculated using the LandScan Global Population Database (2007). For LandScan, sub-national census counts for periods around 2007 were distributed within a 1 km resolution grid based on likelihood coefficients generated from proximity to roads, slope, land cover, night-time lights, and other data-sets. It is the most spatially detailed global population data-set available.

Rangelands in arid zones of Africa cover 6.9 million km^2 (23%) and 118.5 million people (14% of Africa's population). Some 2.9 million km^2 (9.6%) of African arid lands is part of the watershed of a large dam (according to the database of Mulligan, Fisher, et al., 2011, of concrete dams with wall 15 m or higher in 2010) used for water supply or hydropower. Much of this watershed area is subject to very high aridity (i.e. a very low aridity index, shown in blue). Finally, population concentrations of 10 persons/km^2 or more cover 6.9 million km^2 (43% of the African dryland area) and incorporate some 118.5 million people (14% of Africa's total).

Climate change in Africa's drylands

Climate change will impact countries directly as their annual total rainfall and mean temperature change (alongside myriad changes in timing, seasonality and intensity of weather) but also indirectly as climate change impacts neighbouring or even far-away areas which form part of their food, food-water and energy supply chains. Here we examine both of these effects with reference to African drylands in recognition that African drylands provide significant inputs to the food, food-water and energy supply of the Middle East and North Africa region today. Moreover, significant foreign direct investments in African drylands promise even deeper connectivity between food-water-energy demand from unproductive drylands in high-income countries and supplies from more productive drylands in lower-income countries. We focus on changes in precipitation, since this variable has significant direct impacts on food-water. The effects of temperature change are also considered, but given the uncertainties involved and the scale of this study a full analysis of crop-specific growth and development under multi-variable climate variability and change is not possible.

Climate change in Africa's drylands, by basin

WaterWorld is used to apply an SRES A2a 2050s scenario using an ensemble of five GCMs (Van Soesbergen, 2011). The GCMs chosen are CCCMA_CGCM31, CSIRO_MK30, IPSL_CM4, MPI_ECHAM5 and UKMO_HADCM3, all of which are well respected. The A2a scenario is at the higher end of the SRES emissions scenarios, projecting more than a 3°C rise globally by 2100. We examine projected change in annual mean temperature and annual total precipitation for Africa's drylands by basin in Figure A4 and by country in Figure A5. By basin we see that, under this scenario, dryland temperature is expected to increase throughout the region by 1–3°C, with the greatest warming in the dryland areas of basins of the "North Interior" and the "South Interior". Rainfall increases in most African dryland basins but decreases in some. Significant increases are observed throughout the drylands of the "North Interior" basin and in the "North East Coast" basin, the "Rift Valley" and the "Shebelli and Juba" basin. The Nile basin drylands show a moderate increase in rainfall.

By country (Figure A5), warming is greatest in the dryland areas occupied by Algeria, Mali and Botswana. Precipitation is projected to increase the most in dryland areas of Algeria, Mauritania, Ethiopia, Kenya and Somalia and to decrease the most in dryland areas of Mozambique, Angola, Nigeria and Cameroon.

For Africa, there is a weak relationship between pixel-level long-term mean annual temperature according to WorldClim and NASA MODIS (Moderate Resolution Imaging Spectrometer)–estimated long-term total annual actual evapotranspiration (ET) from the satellite climatology assembled by Mulligan (2011). In these data (not shown) higher temperatures do correspond spatially with higher actual ET, and on average an increase in temperature of 1°C leads to an increase in ET of 19 mm. This means that warming of 1–3°C could lead to increases in ET of up to 60 mm/y at the most. This is significantly less than the projected increases in rainfall for much of Africa, so from now on we focus on the rainfall outcomes only. Ideally the WaterWorld water balance model would be run for the whole of Africa to examine the impact of climate change on water balance (including effects on ET, fog inputs and snowmelt). Since this is not possible at the continental scale and rainfall change is clearly the dominant hydrological signal in climate change (compared with ET), we move forward examining rainfall only.

GCM uncertainty

Though an ensemble of five GCMs was used to help manage the significant uncertainty between different GCMs, it should be understood that the results below are highly dependent on the climate scenario and GCMs used and represent the central tendency expected for change. Results from individual GCMs may depart significantly from the central tendency shown here, and different climate scenarios project different rainfall futures. Rainfall projections of GCMs are particularly uncertain in magnitude, spatial pattern and direction of change (see Buytaert et al., 2010; Mulligan, Fisher et al., 2011).

Rainfall change impacting African drylands directly

Here we examine projected changes in rainfall over the key food (croplands and pastures), water and energy (watersheds of dams) geographies of dryland Africa. As shown in Figure A6(a), for cropland areas in drylands, the overall rainfall change is an increase of +59 mm/y (which is sufficient to offset potential higher ET under climate change, possibly resulting in no net change in water balance). Geographically, 90% of dryland croplands show an increase in rainfall, with only 9% showing a decrease. For dryland rangelands (Figure A6(b)), the mean change in rainfall is +100 mm/y, with increases over 95% of dryland rangelands and decreases over the remaining 5%. Examining populated areas (defined as more than 10 people per km^2), the mean change continues to be an increase in rainfall that is sufficient to offset the greater ET (Figure A6(c)). Rainfall increases over 91% of the populated dryland area (affecting 100 million people) and decreases over the remaining 9% (affecting 17 million people).

Rainfall change in the watersheds of dams affects hydropower and water resources for domestic, industrial and agricultural use. For dam watersheds with areas of their catchment in the African drylands, the mean rainfall change for each watershed is shown in Figure A6(d). The mean change in rainfall is +66 mm/y, with 92% of dammed areas showing rainfall increases upstream and 8% showing decreases upstream. For the parts of watersheds within drylands, the mean rainfall change is +81 mm, with 95% of the watershed area showing more rainfall and 5% less (Figure A6(e)). We can thus expect climate change to increase the climatic productivity of cropland and rangelands in African drylands and to generate more water in the watersheds of dams in these same regions – all in all a climatic "net positive" for water, food and energy in these regions, with potential benefits locally, downstream and along the supply chain. Whether the increase in climatic productivity is realized in yields will depend very much on investment conditions, soil conditions and land management and on a wealth of other biophysical and socio-economic factors. We now compare this African dryland situation with the climate change projections for all areas in Africa to understand to what extent drylands mirror the overall continental projections for change.

Rainfall change in all of Africa

The impact of climate change in drylands reflects the overall pattern for Africa of increasing rainfall. Across all cropland areas, the mean projected change in rainfall is +32 mm/y, with increases over 64% of croplands and decreases over 36% (Figure A7(a)). Over all African rangelands, the mean rainfall change is +64 mm, with increases over 64% of rangelands and decreases over 36%. For populated areas, the mean rainfall change is +32 mm/y, with increases over 70% of areas (affecting 590 million people directly) and decreases over 29% of areas (affecting 270 million people directly).

Rainfall change in the watersheds of Africa's dams indicates an overall wetting of +33 mm/y, with increases over 85% of the area and decreases over 15%. Over the largest basins, rainfall increases are small – on the order of the expected ET changes, and thus likely to offset those changes. The greatest rainfall increases in dam watersheds are expected for the Akasombo dam system in Ghana, whilst for some dams in west-central Africa and south-eastern Africa, decreases in rainfall are expected. Overall, we can expect African dams to be receiving more water annually under climate change, and if the dams are engineered to cope with this, the outcome should be an improvement in their productivity. The seasonality and intensity of the increased rainfall will be important, especially if changes in those properties mean that the watersheds of dams produce greater sediment inputs to reservoirs that could reduce reservoir capacity, damage turbines and thus offset these projected positive outcomes.

Alternative scenarios

Given that these rainfall increases differ from some of the previously published research that used different GCMs, scenarios and definitions of dryland (e.g. Collier, Conway, & Venables, 2008; Cooper et al., 2008), here we examine some other GCM ensembles and emissions scenarios to understand the extent to which this five-GCM ensemble mean is anomalous. Figure A8(a) shows the number of GCMs (from an ensemble of 17) agreeing with an increase in rainfall (>0) for the SRES A2a scenario. For the driest of the drylands, all models agree on wetting; for the wetter parts 9–14 out of 17 agree. For the A1b scenario (Figure A8(b)) the uncertainty is greater, with most GCMs agreeing on wetting in the hyper-arid regions and in the Horn of Africa but all agreeing on drying in coastal North Africa and most (10–23) agreeing on drying in south-west Africa. Depending on the emissions scenario used, rainfall is projected either to increase or to decrease in the most populous of the drylands.

Propagation of rainfall change in African drylands along food-water and energy supply chains

WaterWorld's supply chain module is used to examine the 'teleconnections' between climate change in Africa's drylands and potential impacts on water-relevant commodity flows produced in these regions. A teleconnection is a relationship between a variable or variables over large distances, typically thousands of kilometres. The WaterWorld supply chain module uses the UN's Comtrade database (UN, 2013) of export values (in USD) for the years 2007–2011 to map the key flows of agricultural products from dryland Africa as a means of highlighting international dependencies on agriculture in these areas. We calculate the proportion of national-scale exports that could derive from Africa's drylands according to the distribution of associated land. For Comtrade's "all exports" we use the entire national territory, but for specific commodities we use the distributions of associated land covers that support the different commodity flows. We use Ramankutty et al. (2008) (cropland) for Comtrade "crops"; Ramankutty et al. (2008) (pasture) for Comtrade "meat products" and "dairy products"; cropland and pasture for Comtrade "food products"; Monfreda (2008) (cereals) for Comtrade "cereal grain"; and Mulligan, Fisher, et al. (2011) dam watersheds (but also WaterWorld's surface mines and oil and gas) layers for Comtrade "electricity". In this study these areas are masked for drylands, herein defined. The export values are rescaled according to projected annual total change in rainfall in drylands under these land covers. Thus, the figures presented here are the baseline rainfall totals and scenario changes in rainfall supporting the commodity flows. If the commodity connections between countries remain broadly similar to today's, these changes in rainfall are likely to affect those countries importing from drylands, either through change in supply (and thus price) or through the need to reconfigure imports to capture other supplies. It is important to recognize that the focus here is on changes in climate (rainfall) supporting trade flows, not changes in trade itself. We assume that both the pattern and magnitude of change remain the same as today's and focus only on the climate change–related flow changes as an indicator of the propagation of climate risk through commodity supply chains.

No export data were available in Comtrade for the following African countries: Angola, Benin, Botswana, Cameroon, Chad, Comoros, Rep. Congo, Dem. Rep. Congo, Equatorial Guinea, Eritrea, Gabon, Guinea-Bissau, Guinea, Lesotho, Liberia, Libya, Mali, Morocco, Mozambique, Sao Tome and Principe, Seychelles, Sierra Leone, Somalia, Swaziland and Western Sahara. The results here are based on Comtrade export data for Algeria, Burkina Faso, Burundi, Central African Republic, Cote d'Ivoire, Djibouti, Egypt, Arab Rep., Ethiopia, The Gambia, Ghana, Italy, Kenya, Madagascar, Malawi, Mauritania, Namibia, Niger, Nigeria, Rwanda, Senegal, South Africa, Spain, Sudan, Tanzania, Togo, Tunisia, Uganda, Rep. Yemen, Zambia and Zimbabwe. We first examine the baseline (current) rainfall falling on dryland areas in Africa and thus supporting current supply chains for commodities grown in those areas. This gives an idea of the key beneficiary countries of current rainfall in Africa's drylands. We then examine what kinds of change those beneficiaries might expect to see under climate change. We could have calculated green water (ET) for the baseline, but since we cannot calculate this value for the scenario, we focus instead on total rainfall as the embedded quantity available to support production and we consider this rainfall and rainfall change to flow along supply chains. There are clearly complexities in the potential translation of rainfall to biological productivity and to commodity production, which relies on investment, infrastructure and markets, but nevertheless, with all else equal, changes in rainfall will create or hamper such opportunities.

Baseline supporting rainfall flows from dryland Africa

All commodities (total: 24,646,290 km³ per 5 years)

For all commodities,[1] flows of supporting rainfall from Africa's drylands to other countries range from 49 million m³ per 5 years to 377,246,765 million m³ per 5 years, with key beneficiaries in neighbouring countries of North and West Africa, Southern Europe, the Middle East, South America and Australia (Figure A9).

There are, however, significant differences in the rainfall teleconnections between different commodities as a result of the countries from which particular sink countries source their commodities, the area of dryland in those countries, and its spatial relationship with the source areas for the commodity in question – and of course the rainfall in these dryland source areas (see Figure A10). For food products there are strong supporting flows to Africa and Southern Europe. For crops not elsewhere classified, flows are greatest to Asia and Eastern Europe. For cereal grains, the greatest flows are within Africa and the Middle East.

Dairy and meat are assumed to be produced in the same locations, as defined by the distribution of pastures, but countries to which supporting water flows are greatest differ between these commodity classes because of differences in the African dryland source countries that importing countries source their meat and their milk from. For electricity, supporting rainfall flows are greatest to African and Middle Eastern countries; flows to

China probably represent Chinese investments in electricity production in Africa. Having understood the countries benefitting from rainfall supporting production in Africa's drylands under current conditions, we now move to analyze the likely impacts of climate change on these received benefits.

Change in supporting rainfall flows for all commodities (total: +9,876,102 km³ per 5 years, 40% of baseline rainfall)

In line with the generally higher rainfall projections for Africa's drylands, the net change in supporting rainfall flows for all commodities exported from Africa's drylands is around +9876 billion m³ of water (9876 km³), or 40% of the baseline value, with particularly high increases in supporting rainfall for dryland exports to other countries in Africa and Southern Europe (Figure A11).

As shown in Figure A12 and Table 1, virtually all export destinations have more supporting rainfall under climate change in Africa's drylands. Only some very small territories have less. This reflects the small number and volume of commodities exported to these countries, which allows potential domination by commodities with a small land footprint that just happens to coincide with one of the dryland's limited areas of rainfall decline under climate change.

Discussion and conclusions

African drylands cover a significant proportion of the continent, are home to much of the continent's cropland and rangelands, and support significant local human populations. Climate change will impact Africa. These impacts may serve to increase water availability through increased rainfall. Contrary to popular perception, this could potentially lead to

Table 1. Top 10 countries for increasing or decreasing supporting rainfall flows under climate change along supply chains from Africa's drylands for all commodities. Based on data from UN Comtrade, DESA/UNSD. All data for 2007–2011, in million m³ per 5 years.

Top 10	
Senegal	210,000,000
Rep. Congo	180,000,000
Portugal	180,000,000
Benin	140,000,000
Argentina	130,000,000
Greece	130,000,000
Egypt	130,000,000
Australia	130,000,000
Nigeria	120,000,000
Georgia	110,000,000
Bottom 10	
Tonga	2,400,000
Niue	2,300,000
Bhutan	2,200,000
Wallis Island	1,800,000
Kiribati	140,000
Vatican City	130,000
Palau	42,000
W. Sahara	110
Montserrat	− 180,000
St. Pierre	− 230,000

Africa's becoming a bright spot for future food security locally, and a significant breadbasket providing increased supply to global food-water-energy supply chains. This would be good news for Africa and beyond but is highly sensitive to GCM uncertainty since these water positives are the result of increased rainfall offsetting the enhanced ET arising from warming. GCMs are particularly uncertain when it comes to rainfall projections, so a multi-model ensemble was used to capture and reduce the overall projection uncertainty; but significant uncertainty remains between different emissions scenarios.

The benefits of this increased rainfall will be felt internationally along the supply chains for commodities derived from Africa's drylands and will vary with commodity. The impact of climate change on a particular country is thus the direct impact of climate change within the country in question combined with the impacts of climate change that teleconnect along the commodity supply chains originating in all of the countries from which it sources water-food-energy. The drawbacks of changing rainfall in dryland Africa will be experienced differentially by countries throughout the world because of these supply chain teleconnections. The extent to which a country benefits depends on: (1) the magnitude of the commodity flow; (2) the extent to which it is sourced in drylands in Africa; (3) the geographical footprint from which the commodity is sourced; (4) the overlap of GCM-projected rainfall change with that footprint; and (5) the number of import sources that a country is reliant on and the distribution of these sources relative to projected patterns of climate change.

This is the first study to map Africa's climate future including its impact on water-food and energy both locally and along international supply chains. The exercise shows the relative food-water-energy productivity of African drylands and the extent of local, nearby and also remote teleconnections with this productivity and the rainfall that supports it. It shows how climate change could affect these teleconnections and, under some emissions scenarios, could lead to the emergence of Africa as an increasingly important source of food-water-energy–related commodities. The extent to which this happens will depend on (1) to what extent the projections of increased rainfall are accurate; (2) the accuracy of the commodity distribution maps and export data used to represent current and future distributions of these elements; and (3) whether local and foreign direct investment provide the African agricultural capacity to make use of this extra water and avoid soil and other limits to productivity.

Dryland (arid) Africa faces huge climate uncertainty – as it has always done, but this time it supports more people, more cropland and more cattle. Areas classified as arid and hyper-arid in Africa are home to some 142 million people, 2.7 million km^2 of cropland and 6.9 million km^2 of pasture. Some climate models suggest that these areas will dry significantly, others say they will remain broadly the same, and yet others say they will become wetter. The geographical pattern of projected change is complex. Though all indicate that drylands will experience warming, the uncertainty in annual total rainfall projections is huge, and even greater if we examine rainfall seasonality – which is of course as important as annual total rainfall for food production. In the face of this uncertainty, what are people in African drylands to do? It is clear from this analysis that we really do not know what the climate future holds, so we must focus on being adaptable to a range of possible futures: drying, wetting, drying then wetting, wetting then drying, and so on. The more diverse, efficient, sustainable and connected are the systems for dryland Africa's food, water and energy, and the more integrated (nexus-ed) are its policies for water, energy and food production, the better prepared it will be to adapt to whatever comes. Nobody has a crystal ball laying out the climatic future, so now is the time to focus on being more adaptable.

Acknowledgements

This article draws on analysis with the WaterWorld Policy Support System, which has been developed over many years under a wide range of EU, CGIAR Challenge Programme on Water and Food and other funding sources, though is not directly funded by any of them. The many providers of global data-sets used in WaterWorld and of the Comtrade database are also gratefully acknowledged.

Disclosure statement

No potential conflict of interest was reported by the author.

Supplemental data

Supplemental data for this article can be accessed at http://dx.doi.org/10.1080/07900627.2015.1043046.

Note

1. Full list at https://www.gtap.agecon.purdue.edu/databases/contribute/detailedsector.asp

References

Buytaert, W. W., Vuille, M. M., Dewulf, A. A., Urrutia, R. R., Karmalkar, A. A., & Célleri, R.R. (2010). Uncertainties in climate change projections and regional downscaling in the tropical Andes: implications for water resources management. *Hydrology and Earth System Sciences*, *14*, 1247–1258. doi:10.5194/hess-14-1247-2010

Collier, P., Conway, G., & Venables, T. (2008). Climate change and Africa. *Oxford Review of Economic Policy*, *24*, 337–353. doi:10.1093/oxrep/grn019

Cooper, P. J. M., Dimes, J., Rao, K. P. C., Shapiro, B., Shiferaw, B., & Twomlow, S. (2008). Coping better with current climatic variability in the rain-fed farming systems of sub-Saharan Africa: an essential first step in adapting to future climate change? *Agriculture, Ecosystems & Environment*, *126*, 24–35. doi:10.1016/j.agee.2008.01.007

Hijmans, R. J., Cameron, S. E., Parra, J. L., Jones, P. G., & Jarvis, A. (2005). Very high resolution interpolated climate surfaces for global land areas. *International Journal of Climatology*, *25*, 1965–1978. doi:10.1002/joc.1276

LandScanTM Global Population Database. (2007). Oak Ridge, TN: Oak Ridge National Laboratory. Retrieved from http://www.ornl.gov/landscan

Monfreda, et al. (2008). Farming the planet: 2. Geographic distribution of crop areas, yields, physiological types, and net primary production in the year 2000. *Global Biogeochemical Cycles*, *22*, GB1022. doi:10.1029/2007GB002947

Mulligan, M. (2009). *Global mean dry matter productivity based on SPOT-VGT (1998–2008)*. Retrieved from http://www.ambiotek.com/dmp

Mulligan, M. (2006). *MODIS MOD35 pan-tropical cloud climatology*. Version 1. September 2006. Retrieved from http://www.ambiotek.com/clouds

Mulligan, M. (2011). *SimTerra: A consistent global gridded database of environmental properties for spatial modelling*. Retrieved from http://www.policysupport.org/simterra [Mean ET 2000–2010 from Mu, Q., M. Zhao, S.W. Running (2011) Improvements to a MODIS Global Terrestrial Evapotranspiration Algorithm.Remote Sensing of Environment, Volume 115, pages 1781–1800 (doi:10.1016/j.rse.2011.02.019)].

Mulligan, M. (2013). WaterWorld: a self-parameterising, physically-based model for application in data-poor but problem-rich environments globally. *Hydrology Research*. doi:10.2166/nh.2012.217

Mulligan, M., Fisher, M., Sharma, B., Xu, Z. X., Ringler, C., Mahé, G., ... Ahmad, M. U. D. (2011). The nature and impact of climate change in the Challenge Program on Water and Food (CPWF) basins. *Water International, 1941–1707, 36*, 96–124.

Nakicenovic, N. (Ed.). (2000). *Special report on emissions scenarios*. Cambridge, MA: Cambridge University Press. 599 pp.

Ramankutty et al.. (2008). Farming the planet: 1. Geographic distribution of global agricultural lands in the year 2000. *Global Biogeochemical Cycles*, *22*, GB1003. doi:10.1029/2007GB002952

Ramirez, J., & Jarvis, A. (2010). *Disaggregation of Global Circulation Model outputs*. Decision and policy analysis working paper no. 2. Retrieved from http://www.ccafs-climate.org/media/ccafs_climate/docs/Disaggregation-WP-02.pdf

Trabucco, A., & Zomer, R. J. (2009). *Global Aridity Index (Global-Aridity) and Global Potential Evapo-Transpiration (Global-PET) Geospatial Database*. CGIAR Consortium for Spatial Information. Published online, available from the CGIAR-CSI GeoPortal at: http://www.csi.cgiar.org/

United Nations Environment Programme. (1997). *World atlas of desertification 2ED*. London: UNEP.

United Nations Commodity Trade Statistics Database, Department of Economic and Social Affairs/ Statistics Division. (2013). COMTRADE database. Retrieved from http://comtrade.un.org/db/

Van Soesbergen, A. (2011). Global 1 km gridded multimodel climate dataset based on CIAT statistically downscaled data. Retrieved from http://gisweb.ciat.cgiar.org/GCMPage/download_sres.html

Towards a water–energy–food nexus policy: realizing the blue and green virtual water of agriculture in Jordan

Samer Talozi[a], Yasmeen Al Sakaji[b] and Amelia Altz-Stamm[c]

[a]Civil Engineering Department, Jordan University of Science and Technology, Irbid;
[b]Bioenvironmental and Irrigation Engineering, Jordan University of Science and Technology, Irbid;
[c]Lyndon B. Johnson School of Public Affairs, University of Texas at Austin, USA

Virtual water is an important addendum to how we view a country's water resources. This study examines the virtual water embedded in Jordan's agricultural produce and its impact on future water–energy–food policies. Blue and green virtual waters are calculated from data on rainfall, crop patterns, yields, and water requirements at the district level. Results highlight the advantages of blue water usage in the Jordan Valley and of harnessing more available green water in the Highlands, with both displaying low energy impact. Results also emphasize the high groundwater usage and energy footprint in the Desert regions, signalling a need to rein in groundwater extraction and take advantage of solar power.

Introduction

Virtual water (VW) refers to the water used in the production processes of a commodity, good or service (Hoekstra & Hung, 2002). Since its inception, the VW concept has been used in various research topics and at varying scales. A growing number of researchers are using the VW concept as a tool in water policy, management and/or global trade analysis, and this requires accurate virtual water estimates. Studies have covered scales ranging from the basin (Aldaya, Martínez-Santos, & Llamas, 2010; Dietzenbacher & Velázquez, 2007; Montesinos, Camacho, Campos, & Rodríguez-Díaz, 2011), to the national (Guan & Hubacek, 2007; Roth & Warner, 2008; Wichelns, 2001; Zhang, Yang, Shi, Zehnder, & Abbaspour, 2011), to the global (Fader, Rost, Müller, Bondeau, & Gerten, 2010; Yang, Wang, Abbaspour, & Zehnder, 2006).

Notwithstanding the widespread use of this concept, a number of researchers have voiced concern about its method of usage and adoption into policy. Wichelns (2010) argues that the VW perspective cannot be used alone as a criterion for selecting optimal policies. Similarly, Aldaya et al. (2010) explain that despite the ability of the VW concept to offer a framework for the identification of potential optimal policy alternatives, it still requires additional complementary balancing tools such as risk diversification and inclusion of labour, social and economic factors.

With regard to agriculture and food production as addressed in this study, virtual water has been extensively studied because agriculture accounts for about 70% of human blue water withdrawals – surface and groundwater (Molden et al., 2007). There is support for the potential value of this concept in irrigation planning, management and policy, provided that two major conditions are in place. First, an accurate calculation of VW has to be obtained that takes into account the spatio-temporal environmental variability and different combinations of water types: blue, green and treated wastewater (Antonelli, Roson, & Sartori, 2012). Second, VW should be considered in conjunction with other relevant social, political and economic factors based on the context of the geographical area in question (Daniels, Lenzen, & Kenway, 2011). Previous attempts to compute VW for nations have relied on average climatic values and/or were generated using global models. As a result, when downscaled to the country level, current available VW data are neither accurate nor properly classified as blue or green water.

Chapagain and Hoekstra (2004) calculated the virtual water content for 175 crops and 210 countries, including Jordan, but their approach had a number of limitations. First, they used country averages of climatic variables. Second, they assumed no water limitations, which is not the case in Jordan. Third, they did not distinguish between blue and green water usage. Siebert and Döll (2010) computed blue and green virtual water for 26 crops using a global crop water model with data from 1998–2002, but because of the uncertainties and limitations in their input data, model parameterization and model structure, it was only useful for global-scale assessments. Mourad, Gaese, and Jabarin (2010) calculated virtual water in the Jordan Valley using more accurate data and taking into consideration the quota used for water distribution, but their study was limited by its sole focus on fruit trees in the Jordan Valley and its not taking into account the distinction between blue and green water.

Other studies have shown that Jordan, faced with a huge water deficit, has externalized much of its virtual water use and in the process has become more dependent on other nations (AlAyyash, Al-Adamat, & Al-Meshan, 2012; Chapagain & Hoekstra, 2003). Increasing inter-sectoral competition for water, the need to feed and provide drinking water for an ever-growing population and waves of refugees, water scarcity and the potential adverse impacts of climate change are further causes for concern in Jordan (Mercy Corps, 2014). In addition, the estimated portion of non-revenue water in Jordan is 50%, of which around 70% is due to illegal usage (Ministry of Water and Irrigation, 2009). This context is emblematic of Jordan's weak governance of water resources (Ministry of Water and Irrigation, 2009), and thus a detailed and accurate look at the current water allocation practices in Jordan is warranted. This is particularly true in the agricultural sector, which consumed around 55% of the total available water resources in Jordan in 2010 (Ministry of Water and Irrigation, 2013), and the VW concept provides an appropriate framework to offer more optimal management practices.

Distinguishing between the blue and green virtual water content of crops is also important from a sustainability point of view. Blue water is commonly defined as surface and groundwater. For green water, this study uses the FAO's definition of effective rainfall, which is essentially the usable rainfall or the amount of rainfall during the growing season that can be used to meet crop water requirements. In turn, ineffective rainfall is the amount that is lost as surface runoff, deep percolation, or moisture remaining in the soil after crop harvesting and not used in the subsequent season (Dastane, 1978).

This distinction between blue and green water is especially of note in a country like Jordan, where over the past three decades there has been a dramatic shift from using mainly green water in agriculture to using mainly blue water, particularly the country's diminishing groundwater resources. By separating out blue and green virtual water instead of combining the two, the concept of the water footprint or impact of nations and individuals is likely to gain some much-needed clarity (Hoekstra & Chapagain, 2008; Yang et al., 2006).

In addition, treated wastewater is still considered part of blue virtual water and is not to be confused with grey virtual water, which is the polluted water left over from the production of a product (Berger & Finkbeiner, 2010). This research will highlight treated wastewater as a unique type of blue water with the label 'recycled blue water'.

In this work, the VW embedded in agricultural production in Jordan will be more accurately estimated by taking into account the diversity of the agricultural sector and the placement of VW within this sector. Our analysis includes a distinction between blue and green virtual water and special recognition for recycled blue water's place in blue virtual water. It also takes into account the spatial variability of rainfall within Jordan's three unique climatic zones. And to fully furnish the water–energy–food nexus, this study offers a spatial distribution of energy use in the irrigation sector, something not previously seen, although other studies have discussed overall water–energy interactions in Jordan (Hellegers, Zilberman, Steduto, & McCornick, 2008; McCornick, Awulachew, & Abebe, 2008; Meisen & Tatum, 2011). Finally, there is a discussion of the practical implications of the results and the potential for more sustainable and energy-efficient policies in the agricultural sector. This detailed examination of VW at the national level is intended to enable the realization of the water–energy–food nexus and to shed light on ways to improve water and energy resources management at the national level in Jordan.

Study area

Jordan is divided into three major agricultural climatic zones: the Jordan Valley, the Desert (sometimes referred to as the Badia) and the Highlands (Figure 1). There are 12 governorates and 73 districts, with this research being carried out at the district level thanks to the availability of a detailed district-level survey conducted by Jordan's Ministry of Agriculture in 2012. Each of these three agricultural zones has its own characteristics with regard to cropping patterns, water resources and socio-economic conditions. Variations, however, exist not only among the zones but also within them. These variations are related to the source of irrigation water and the amount of energy used per unit volume of irrigation water.

The Jordan Valley is divided into four subzones: North, Middle, Karama (also known as South Shouneh) and South (Figure 2). Agriculture in the Jordan Valley is primarily irrigated due to the limited rainfall and high evapotranspiration rates. Water resources vary according to its subzones (Table 1). Fresh surface water originating in the Yarmouk and Jordan Rivers is more available in the North, with surface water also coming from a number of side valleys that flow from east to west in the Jordan Valley. The Middle and Karama subzones, on the other hand, use treated wastewater (recycled blue water) that originates at the King Talal Dam and flows into the Jordan Valley through the Zarqa River. There is also some use of brackish groundwater in Karama (Abu-Sharar, 2006). Farms in the South subzone use surface water originating from fresh springs or mountain runoff. Throughout the Jordan Valley from the North to Karama, the King Abdullah Canal serves as the main irrigation-water conveyance system.

Limited green water (effective rainfall) is available from the North to Karama during the winter. The North is the largest beneficiary of rainfall. In the Middle, rainfall does occur but, due to the widespread use of greenhouses, it is not taken advantage of in the same manner. And in Karama, rainfall occurs less frequently.

The crops grown within the Jordan Valley differ between subzones as well (Table 1). In the North, farmland is largely dominated by citrus trees, but there are also open-field vegetable farms, interspersed more frequently as one moves south. The Middle is almost all

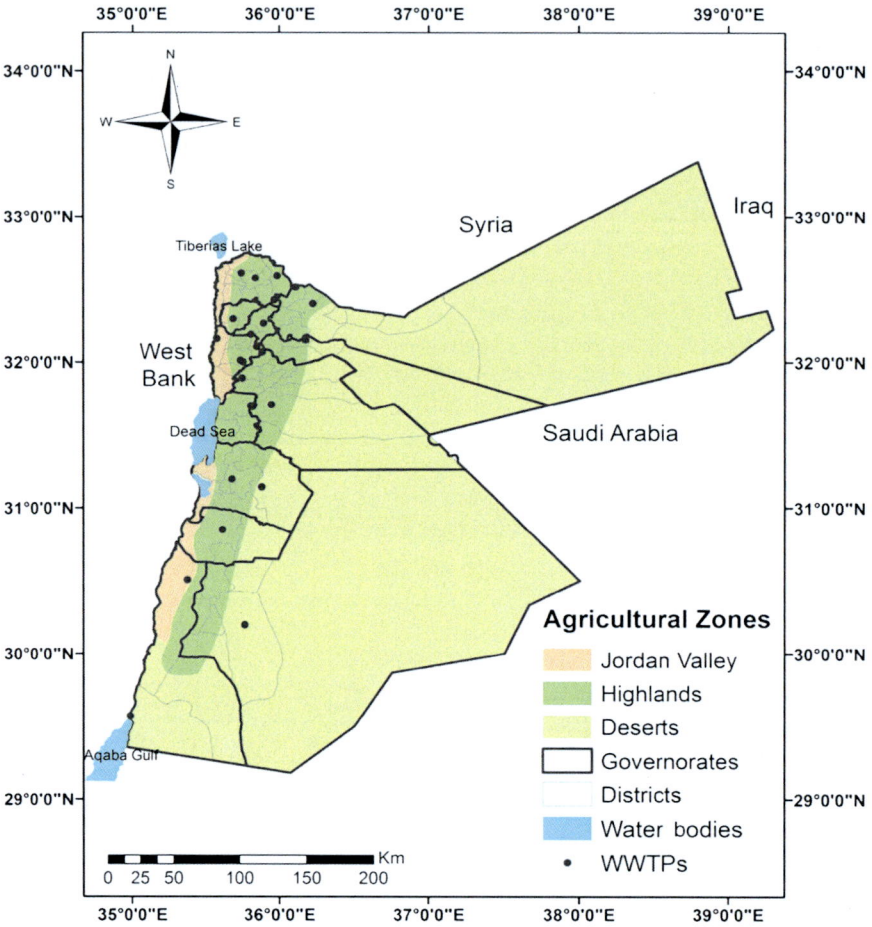

Figure 1. Map of Jordan showing the three agricultural zones, districts and the distribution of wastewater treatment plants (WWTPs).

vegetable crops, much of which are grown in greenhouses. Karama is also predominately vegetable crops, mainly open-field, but date palm trees have become more numerous. It should be noted that in subsequent categorizations and calculations, date palms are included in the amounts for fruit trees. Farms in the South grow mainly open-field vegetable crops. In all zones, most farmers sell their produce in the local market while a smaller portion of farmers with greater capabilities export produce to neighbouring or European countries.

Energy use in the Jordan Valley also varies from North to South. Surface water reaches the King Abdullah Canal by gravity, to a large extent, although some pumping is required through the distribution system and where water is pumped to individual farm ponds. On the farm, water is then pumped by farmers (using diesel or electricity) from their holding ponds to their irrigation systems (Venot, Molle, & Hassan, 2007). Where groundwater is used in the Jordan Valley, it is typically brackish water that is pumped by individual farmers into their holding ponds and usually mixed with the surface water (blue water) or treated wastewater (recycled blue water), after which it is pumped into the irrigation systems. Farmers with more financial capacity have small desalination plants to treat the water before using it on their farms. Using groundwater in this way is energy-intensive.

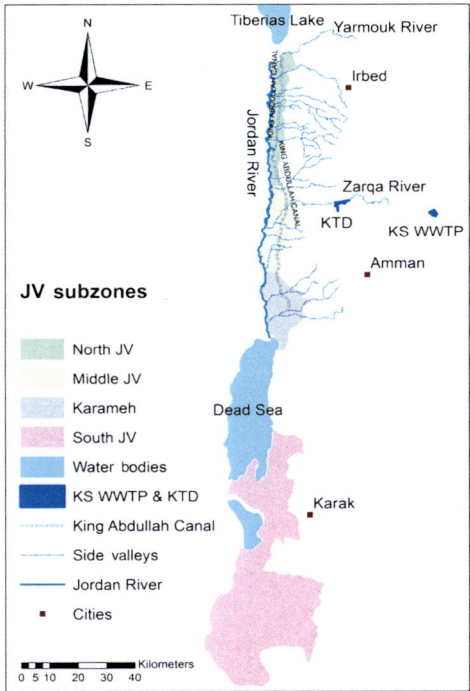

Figure 2.   Map of the Jordan Valley (JV) showing its four agricultural subzones, the Jordan and Yarmouk Rivers, the King Abdullah Canal, the King Talal Dam (KTD), and the Khirbat as-Samra wastewater treatment plant (KS WWTP).

Farms in the Jordan Valley are all of similar size (about 35 *dunums* per farm unit; 1 *dunum* is 1000 m^2), with some farmers operating only part of a farm unit and others operating as many as 20 farm units. Farm owners are almost all Jordanian, some originating from valley villages and others from cities outside the valley such as Irbid, Amman and Karak. There are also Egyptian and Pakistani farmers who rent and farm much of the land in the Jordan Valley. Labourers are predominately Egyptian, but there are also Jordanian, Syrian and Pakistani labourers.

In the Desert zone, agriculture depends heavily on groundwater from both renewable and non-renewable aquifers. Rainwater is very limited and rarely utilized, with the exception of its use for field crops. The main crops grown in this area are vegetables and field crops, with a number of fruit trees in certain locations as well. Energy use is very high in this zone due to the increasing depth of groundwater, requiring higher-intensity pumping to retrieve the water. The impact of this kind of agriculture on water resources, soil and ecosystems has been devastating for this zone over the past two decades (Salameh, 2008; Venot & Molle, 2008). Agriculture in this zone relies mainly on foreign labour (Al-Karablieh, 2012), with most farming in the form of larger-scale commercial operations that export their produce.

In the Highlands, limited blue water is used. Agriculture relies mostly on green water in the form of rainfall. Energy use is thus very limited, with no pumping of water to the farm required. The major crops in this area are olives, grapes, figs and some stone-fruit trees that are mainly rainfed. Winter and summer vegetables are also grown, and both are mainly rainfed. Agricultural produce in this zone is mainly consumed within Jordan and is tightly attached to

165

Table 1. Summary of the cropping patterns and water resources in the four subzones of the Jordan Valley.

Subzone	Water resources	Cropping pattern
North	➢ Surface water from the Jordan and Yarmouk Rivers ➢ Limited use of treated wastewater from the King Talal Dam ➢ Rainfall	➢ Citrus trees ➢ Vegetables (open-field)
Middle	➢ Treated wastewater from the King Talal Dam ➢ Limited rainfall	➢ Vegetables (mainly greenhouse but also open-field)
Karama (including South Shouneh)	➢ Treated wastewater from the King Talal Dam ➢ Limited rainfall ➢ Brackish groundwater	➢ Vegetables (open-field) ➢ Date palms
South	➢ Surface water from fresh springs and side valleys	➢ Vegetables

the social fabric of towns and villages where it occurs. Farms are of small-to-medium size, are more family-based, and rely much less on outside labour (Al-Karablieh, 2012).

By distinguishing between these three agricultural climatic zones in Jordan, it can be seen that their use of green and blue virtual waters, as well as their energy usage, differ (Table 2). The Jordan Valley and the Desert use primarily blue water, with only a limited amount of green water. Highlands agricultural production uses almost all green water but is consuming an increasing volume of treated wastewater (recycled blue water) from a number of smaller wastewater treatment plants (Figure 1). Energy usage is lowest in the Highlands, with more moderate usage in the Jordan Valley and extreme usage in the Desert.

Methodology

Agricultural production

The 2012 agricultural survey covers all 73 of Jordan's districts and takes into account a total of 23 fruit, 25 vegetable and 10 field crops. The amount of production in tonnes per

Table 2. Landscape of green virtual water, blue virtual water, and energy use in the three agricultural climatic zones in Jordan.

Agricultural zone	Green water	Blue water	Energy use
Jordan Valley	Limited	Predominant	Moderate
Highlands	Predominant	Limited	Limited
Desert	Limited	Predominant	Intensive

unit area (*dunum*) is also included in the survey. Furthermore, the survey distinguishes between irrigated and rainfed areas, summer and winter vegetable crops, and open-field and greenhouse production. Thanks to access to this detailed national survey, this work represents the finest spatial resolution, is more comprehensive with regard to types of crops, and covers the entire area of Jordan.

Irrigation crop water requirements

This study relies on analyses from recent studies by the Deutsche Gesellschaft für Internationale Zusammenarbeit (GIZ) focusing on the Azraq Basin (Demilecamps & Sartawi, 2010) and the Jordan Valley (Abdel Jabbar & Sobh, 2012). The crop water requirement values produced for the different agricultural zones and crops in Jordan are subsequently used to calculate the blue virtual water, as explained shortly. These GIZ reports contain more local knowledge and in-depth analysis of microclimates and agricultural practices, and they provide more accurate values than those commonly used and provided by the FAO (Chapagain & Hoekstra, 2004). The GIZ's modified crop coefficient values and evapotranspiration reduction factors take into account local climatic conditions and the types of farming and irrigation methods used (Table 3).

Blue virtual water calculation

The present study estimates the blue virtual water of crops as is without having undergone any processing. The virtual water content in m^3 is calculated as the volume of water used

Table 3. Modifications made by the Deutsche Gesellschaft für Internationale Zusammenarbeit to crop coefficient values and evapotranspiration reduction factors.

Parameter	Parameter definition	Modification rate	Reasoning
K_c initial	Crop coefficient for the initial growth stage	− 25%	Plastic mulch reduces evaporation
K_c mid	Crop coefficient for the middle growth stage	+ 10%	Plastic mulch effect
		− 10%	Wind speed > 4 m/s
		− 10%	Relative humidity > 60%
K_c final	Crop coefficient for the final growth stage	+ 10%	Plastic mulch effect
		− 10%	Wind speed > 4 m/s
		− 10%	Relative humidity > 60%
F_r	ET reduction factor due to drip irrigation	− 5 to − 10%	This reduction range is appropriate for crop canopies at the maturity stage
P_r	ET reduction factor due to planting in a greenhouse	− 50%	Greenhouse reduction in evapotranspiration due to the lower wind speed and radiation and the higher humidity inside the greenhouse in comparison to the open field, despite the relatively higher temperature inside the greenhouse

Source: Abdel Jabbar & Sobh (2012) – detailed study of water requirements for fruit trees and vegetables grown in the Jordan Valley by the Deutsche Gesellschaft für Internationale Zusammenarbeit.

during the entire period of crop growth (m^3/*dunum*) multiplied by the area of that crop (in *dunums*) (Aldaya & Llamas, 2008):

$$VW = \sum_{c=1}^{n} WU_c \times A_c \tag{1}$$

where WU_c is crop water use (m^3/*dunum*), A_c is crop area (*dunums*), c is the crop, and n is the number of crops.

Crop water use (in m^3/*dunum*) is calculated from the accumulated crop evapotranspiration over the complete growing period for each crop:

$$WU_c = \sum_{m=1}^{j} ET_{cm} \tag{2}$$

where ET_{cm} is monthly crop evapotranspiration (mm), m is the month, and j is the total number of months in the growing period.

Monthly crop evapotranspiration is calculated from the crop coefficient and the reference evapotranspiration:

$$ET_{cm} = K_c \times ET_o \tag{3}$$

where K_c is the crop coefficient and ET_o is the reference evapotranspiration (mm).

Green virtual water calculations

As previously mentioned, the green virtual water is calculated from the effective annual rainfall, which can be divided over three seasons in Jordan: winter, spring and fall. Rainfall distribution is not constant throughout Jordan, so the country is further divided into three zones, as distinguished by Freiwan and Kadioglu (2008), each with its own rainfall distribution pattern. The first region consists of the northern Highlands, western Amman, Irbid and the very northern tip of the Jordan Valley, with a rainfall distribution of 63%, 23% and 14% in the winter, spring and fall, respectively. For the second region, which includes the central area of Jordan (Amman), the southern Highlands (Shoubak and Rabba) and the broader northern region of the Jordan Valley (Deir Alla), the rainfall distribution is 63%, 25% and 12% in the winter, spring and fall, respectively. The third region consists essentially of all other locations in the Jordan Valley, the east of the country, and those regions of lower elevation in the Highlands, with a rainfall distribution of 54%, 25% and 21% in winter, spring and fall, respectively.

Thus, green virtual water use is calculated according to the type of crop (vegetables, field crops or fruit trees) and according to when these crops use rainfall (winter, spring and/or fall). All types of crops use rainfall in the spring, for example, but only winter vegetables, wheat, barley, and fruit trees use rainfall in the winter, and only olive trees (out of the fruit trees) use rainfall in the fall. Effective rainfall falling outside of the growing season is not considered.

In order to calculate the green virtual water from rainfall, meteorological analysis of rainfall data in the rainy/winter season of 2011–2012 is used (Ministry of Water and Irrigation, 2012b). Total annual rainfall data are available at the surface-basin level. The weighted-average method is used to calculate the annual rainfall for each district based on the different water catchments in each district:

$$R_D = \frac{(\Sigma A_{BCi} \times R_{Ci})}{100} \qquad (4)$$

where R_D is the weighted average rainfall for the district (mm), R_{Ci} is the average rainfall (mm) in catchment i, and A_{DCi} is the percentage of area of the district in catchment i, with $i = 1, 2, \ldots n$ indexing the catchments contained within the district.

From these calculations the effective rainfall (R_{De}) was computed by using an average value representing the Highlands, where most green water is utilized (Jordan Meteorological Department, 2012):

$$R_{De} = 0.65 \sum_{s=1}^{n} R_D \qquad (5)$$

where R_{De} is the effective rainfall (mm) and R_D the weighted average rainfall for the district (mm); s stands for the season (fall, winter or spring) and n for the number of seasons within a crop's growth period.

GIS modelling

A geospatial model is built which links the rainfed cropping pattern per district with the calculated average annual rainfall per district to calculate green virtual water usage. To start, a geodatabase of the cropping pattern is created in which the cropping pattern is assigned one of three major categories: fruits, vegetables and field crops. Each category is further divided into subcategories based on irrigated versus rainfed agriculture, drip versus surface irrigation, and open-field versus greenhouse agriculture. The district rainfall values are also incorporated into the geodatabase to facilitate the calculation of green virtual water according to crop categories and district.

Calculation of blue virtual water incorporates the spatial water requirements for the different crops into the geodatabase, and information about the volume of treated wastewater (recycled blue water) used in each district is incorporated as well. The irrigated cropping pattern per district is linked to the irrigation crop water requirements in each district to calculate blue virtual water usage.

Blue virtual water verification

The blue virtual water is calculated based on local crop irrigation requirements and the cropping pattern as reported by the agricultural survey. The only way to verify the calculated blue virtual water is by comparing it with the volume of water that was used for irrigation in the same year, as reported by the Ministry of Water and Irrigation (2012b) in its National Water Budget. The volume and distribution of recycled blue water is also obtained from this report.

Virtual water exportation

In order to estimate the amount of virtual water exported from Jordan in the form of fruits and vegetables, the weighted average of virtual water content in fruits and vegetables in m^3/tonne is calculated. Blue and green virtual waters are combined, because the composition of exports (green or blue) is unknown.

Energy footprint

The following calculations are done to arrive at the energy footprint (EFP) of water usage in agriculture. It is essentially the energy used to pump water within a farm, either from a holding pond to the irrigation system, as in the Jordan Valley, or from groundwater wells to the irrigation system, as in the Highlands. The concept of the EFP as discussed in this study does not take into account the energy usage of off-farm utility services.

For each district, information regarding the irrigation water source (surface, ground or a mix), the energy source for water pumping within the farm (electricity, diesel or a mix) and the groundwater depth (where groundwater is used) is compiled. Diesel and electricity are used in both groundwater pumping and surface water pumping settings, with diesel being used when a farm is not close enough to the electrical grid. The price of electricity in the agricultural sector in Jordan is set at JOD 0.06/kWh (National Electric Power Company [NEPCO], 2013), and the average price of diesel in 2012 and 2013 is taken to be JOD 0.60/L (Ministry of Energy and Mineral Resources, 2012, 2013).

From informal surveys conducted with water officials in the Jordan Valley and the Highlands report from Demilecamps and Sartawi (2010), the cost of pumping (in JOD per cubic metre of water) was acquired within different farm settings with varying water and energy sources (electricity and/or diesel).

For areas using electricity, pumping costs are converted to EFP:

$$EFP = \frac{C_s}{T_e} \qquad (6)$$

where EFP is the energy footprint of pumped water (kWh/m^3), C_s is the cost of pumping surface water (JOD/m^3), and T_e is the electricity tariff (JOD/kWh).

For areas using diesel, the diesel use per cubic metre of water is estimated by Equation (7) and then converted to energy use per metre depth of groundwater with Equation (8), with the litres of diesel converted to kWh (one litre of diesel $= 10\,kWh$). Finally, the EFP where groundwater is used is calculated by Equation (9).

$$D = \frac{C_g}{P_d} \qquad (7)$$

$$E = 10 \times D \qquad (8)$$

$$EFP = E \times G_d \qquad (9)$$

Here, D is the volume of diesel used for water pumping (L/m^3), C_g is the cost per volume of pumping groundwater ($[JOD/m^3]/m$), P_d is the price of diesel (JOD/L), E is the energy used for groundwater pumping per metre depth of groundwater ($[kWh/m^3]/m$), 10 is the energy conversion factor from a litre of diesel to kWh, and G_d is groundwater depth (m).

For districts using both electricity and diesel, the percentage of each energy source is taken into account and calculations made accordingly using the above equations.

Finally, the total energy footprint per district (kWh) is calculated by multiplying the EFP (kWh/m^3) by the volume of blue water (m^3) used within each of the 42 districts using blue water. The energy footprints per cubic metre of water and per district are spatially presented using GIS mapping tools.

Results and discussion

The blue and green components of virtual water use are calculated and presented according to their use for fruits, vegetables and field crops in Table 4. Jordan's total virtual water usage is 600,351,187 m³/y, consisting of 386,567,625 m³/y from virtual blue water (57%) and 213,783,562 m³/y from virtual green water (43%). Overall, fruits represent 62% of Jordan's virtual water usage, vegetables 27% and field crops 11%. More specifically, fruits are produced with similar amounts of blue (56%) and green (44%) waters, whereas vegetables are produced with blue water (98%) and field crops are produced with mostly green water (73%).

Figure 3 gives spatial representation to these green and blue virtual water numbers, with Figures 4 and 5 including their destination crops (fruits, vegetables or field crops). The distribution of water usage across Jordan's districts is not even in terms of type of water used (blue or green), volume of that type of water used, or types of crops grown. From a general point of view, there is high usage of blue water in several of the districts in the Desert as well as in all of the Jordan Valley, whereas green water usage is more concentrated in the Highlands along with a few of the Desert districts. High concentrations of blue water are used for vegetable growing in much of the Desert areas, with one district exhibiting largely fruit production. The Jordan Valley's blue water usage is represented in the production of both fruits and vegetables, although with a slight edge for fruits, probably due to the large number of citrus trees in the North and date palm trees in Karama. Green water use, on the other hand, is represented largely in fruits and field crops in the Highlands, with field crops, fruits and vegetables making up portions of the green water usage in the Desert.

In further examining the blue water usage among Jordan's climatic zones, what is most at issue is the difference in water source between the zones. Treated wastewater, or recycled blue water, represents 22% of Jordan's overall blue water resources, 58 MCM/y going to the Jordan Valley and 49 MCM/y going to the Highlands. In the Jordan Valley, the blue water is represented primarily by surface waters, from river water, mountain runoff, or treated wastewater (recycled blue water), with the river water and runoff being renewable resources and far less worrisome in their use for the production of agricultural goods than, for example, non-renewable aquifers. By 2020, the Karama area in the Jordan Valley will receive an additional 15 MCM/y of recycled blue water from South Amman, and by 2035, the North Jordan Valley will receive yet more recycled blue water (25 MCM/y) from the city of Irbid and surrounding smaller towns northwest of Amman (International Resources Group, 2013). The potential for the use of recycled blue water in all regions will only increase in the future due to an increase in population and its generation of wastewater as a result of improved lifestyles, and thus increased collection, treatment and availability of wastewater. See Figure 6 for a view of Jordan's current and future recycled blue water usage and Figure 1 for the locations of current wastewater treatment plants.

Table 4. Blue, green and total virtual water usage for fruits, vegetables and field crops in Jordan.

	Blue virtual water (MCM/y)	Green virtual water (MCM/y)	Total (MCM/y)
Fruits	208	164	372
Vegetables	161	3	164
Field crops	18	47	65
Total	387	214	601

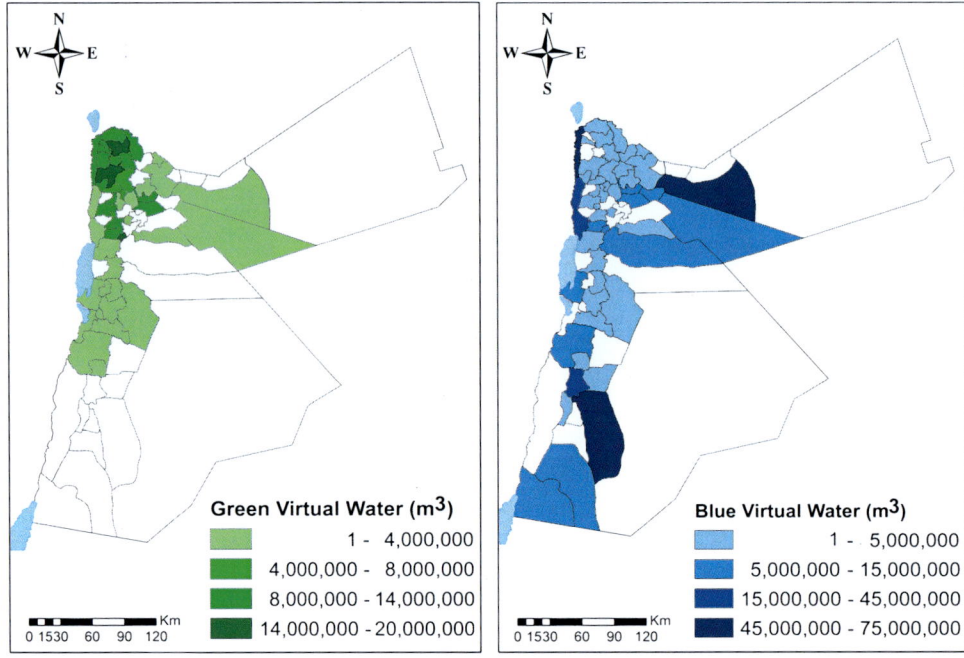

Figure 3.   Spatial representation of green and blue virtual water use by district.

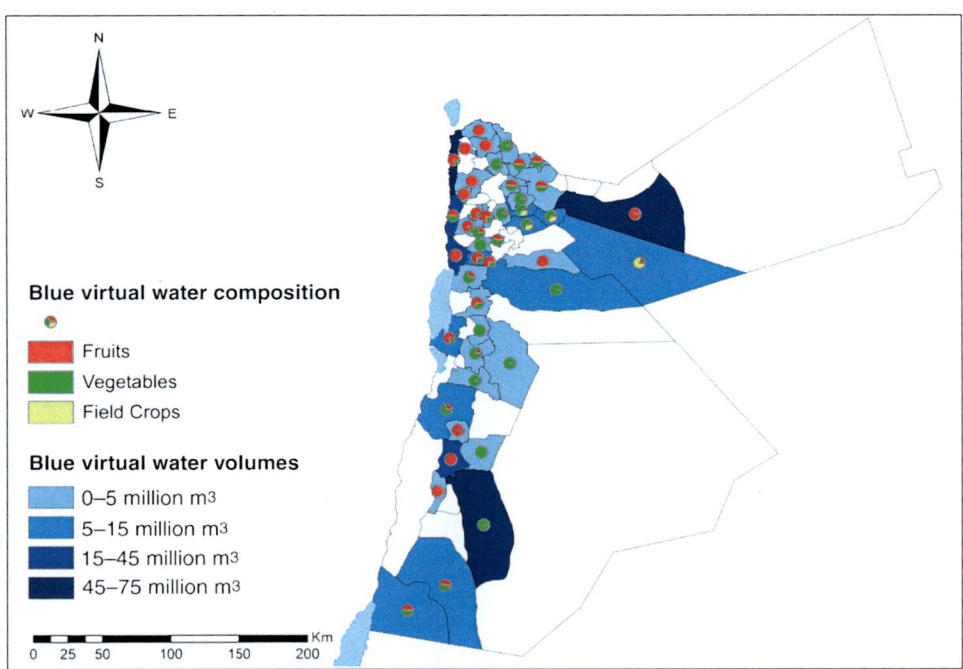

Figure 4.   Blue virtual water volumes computed at the district level, overlaid by pie charts displaying the use of blue virtual water by crop.

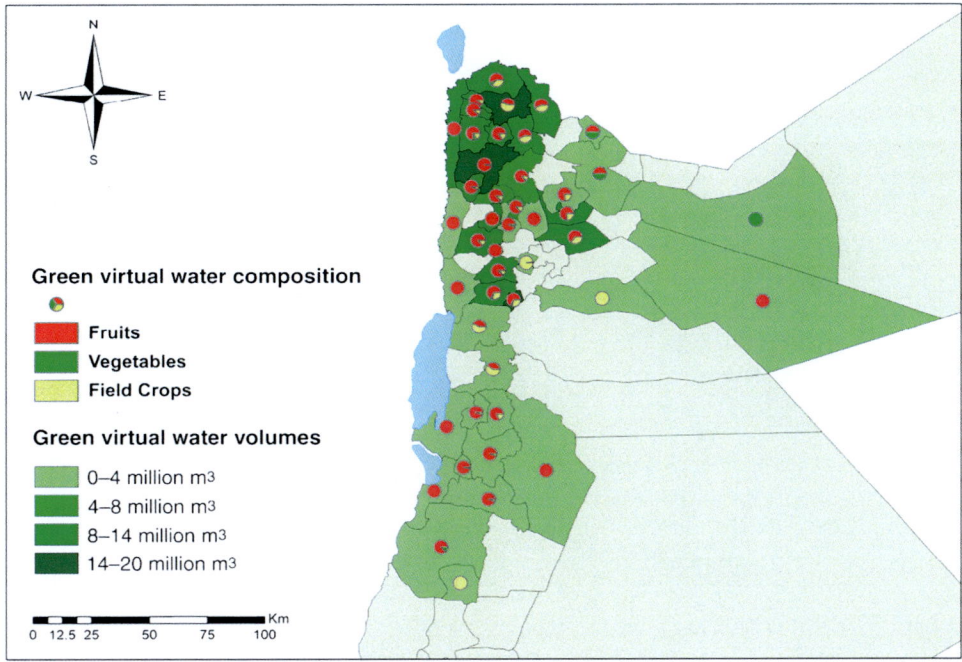

Figure 5. Green virtual water volumes computed at the district level, overlaid by pie charts displaying the use of green virtual water by crop.

In the Desert, on the other hand, the blue water originates solely in groundwater aquifers, both renewable and non-renewable. Over the past two decades, these aquifers have been increasingly strained by the expansion of agriculture in this zone, encouraged by cheap land, cheap water, and the ease of illegally tapping into groundwater resources with few repercussions. Groundwater aquifer safe yields are not being respected, and consequently groundwater levels and quality have deteriorated (Abdel Khaleq, 2011; Abu-Awwad & Blair, 2013). This can be seen, for example, in the Azraq Basin's higher salinity levels as abstraction increases (Figure 7). Up to 40% of Jordan's groundwater system is at risk of depletion by 2030 if current pumping rates continue (Mercy Corps, 2014). To date, there is no use of treated wastewater (recycled blue water) in this zone.

With regard to green water usage and its concentration in the Highlands, despite being a relatively dry to semi-dry environment, this zone is able to sustain its production with rainwater. This makes for a positive impact on Jordan's usage of its precious water resources and represents a variety of agriculture needing very little energy. Unfortunately, the agricultural expansionist trend observed in the Desert is reversed in the Highlands. Due to the further fragmentation of land ownership (Madanat, 2010) and some owners' choosing to no longer farm on their inherited portions of land, many tracts have become urbanized and no longer cultivated (Al-Eisawi, 2012; Parker, 2012). This has prevented the use of green water in many areas that no longer maintain farms.

In verifying the above calculations with Jordan's National Water Budget of 2012, roughly 455 MCM of water was distributed for irrigation purposes, and the calculated total blue virtual water has been found to be 15% less than that amount. This is acceptable considering that there are probably leaks and other seepage of water out of the storage and

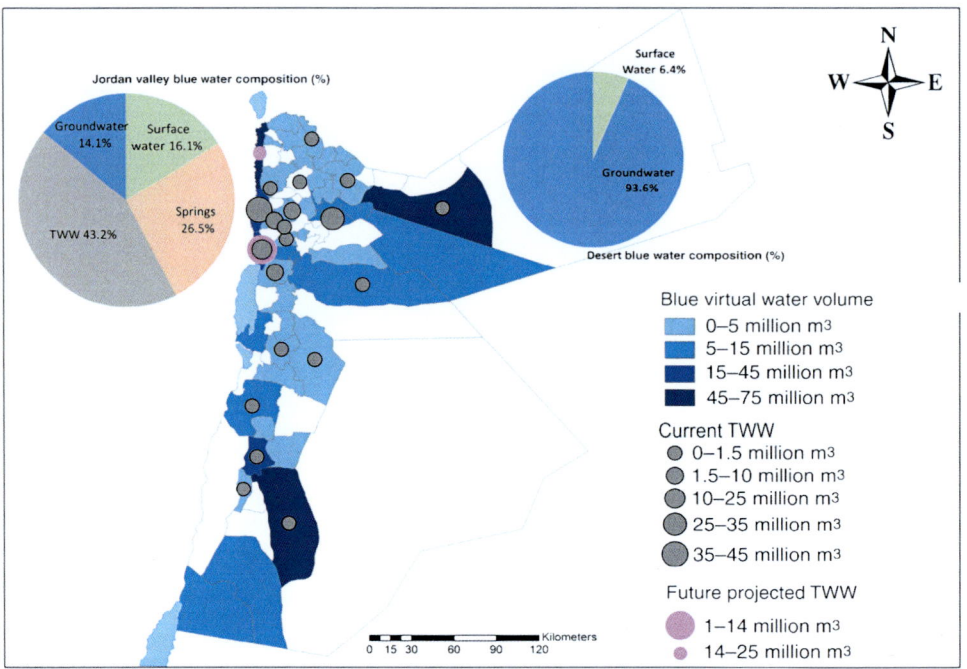

Figure 6.   Volumes of current and future treated wastewater (TWW, recycled blue water) used per district, overlaid onto previous mapping of blue virtual water usage per district.

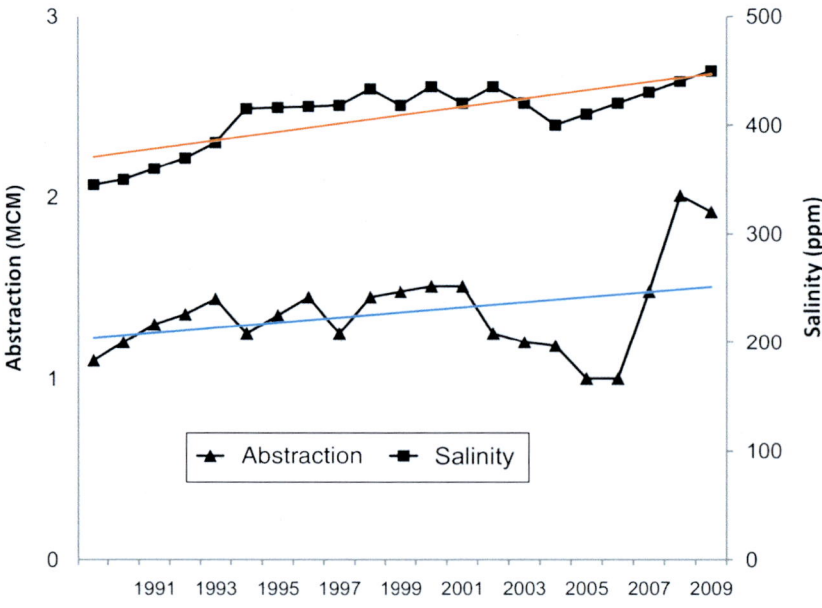

Figure 7.   Relation between abstraction and salinity in a well in Azraq (from the Ministry of Water and Irrigation's Water Information System).

Table 5. Fruits and vegetables exported from Jordan in 2012 in terms of quantity and virtual water content, and fruits and vegetables imported into Jordan in 2012 in terms of quantity.

| | Exported | | Imported |
	Quantity (tonnes)	Virtual water (m^3)*	Quantity (tonnes)
Fruits	109,000	69,666,519	118,000
Vegetables	692,000	136,549,366	112,000
Total	801,000	206,215,885	230,000

Source: Ministry of Agriculture (2012).
*These values were calculated within this study and are not from the Ministry of Agriculture.

conveyance systems throughout Jordan. The Jordan Valley Authority, the government agency responsible for water distribution in the Jordan Valley, also typically operates with the assumption that 10% will be lost in one way or another. These issues could easily explain the 15% difference.

With regard to Jordan's exporting of its virtual water, the average virtual water content of fruits was found to be 639.14 m^3/tonne, and for vegetables 197.33 m^3/tonne, with the virtual water content in the total exported quantity of fruits and vegetables in 2012 thus estimated at roughly 206 MCM (Table 5).

As seen in Tables 5 and 6, out of Jordan's overall production of 1,455,189 tons of fruits and vegetables in 2012, 801,000 tons were directed for export. That means that Jordan exported 55% of its fruit and vegetable production in that year, with the use of 206 MCM of its total virtual water usage. On the other hand, Jordan imported far fewer tonnes of fruits and vegetables than it exported, once again perhaps the opposite of what one would logically expect a water-poor country to do. A similar analysis of the virtual water content of imported fruits and vegetables was not possible because the source of these products, which would impact their virtual water content, is beyond the scope of this study.

As regards the energy portion of the overall nexus concept, Figure 8 presents the energy footprint per unit of water by district and the overall annual energy footprint by district. It is clear that energy usage is especially high and problematic in several of the Desert districts and the lower Highlands districts, precisely the areas where groundwater is being used. On the other hand, in the Jordan Valley and the upper Highlands energy usage is generally lower; this corresponds to their usage of surface waters, treated wastewater (recycled blue water) and/or rainfall.

Table 6. Production of fruits and vegetables in Jordan in 2012.

| | | Production (tonnes) | |
		Blue water	Green water
Fruits	Highlands and Desert	198,985	236,445
	Jordan Valley	158,292	
Vegetables	Highlands and Desert	479,437	64,830
	Jordan Valley	317,200	
	Subtotal	1,153,914	301,275
	Total	1,455,189	

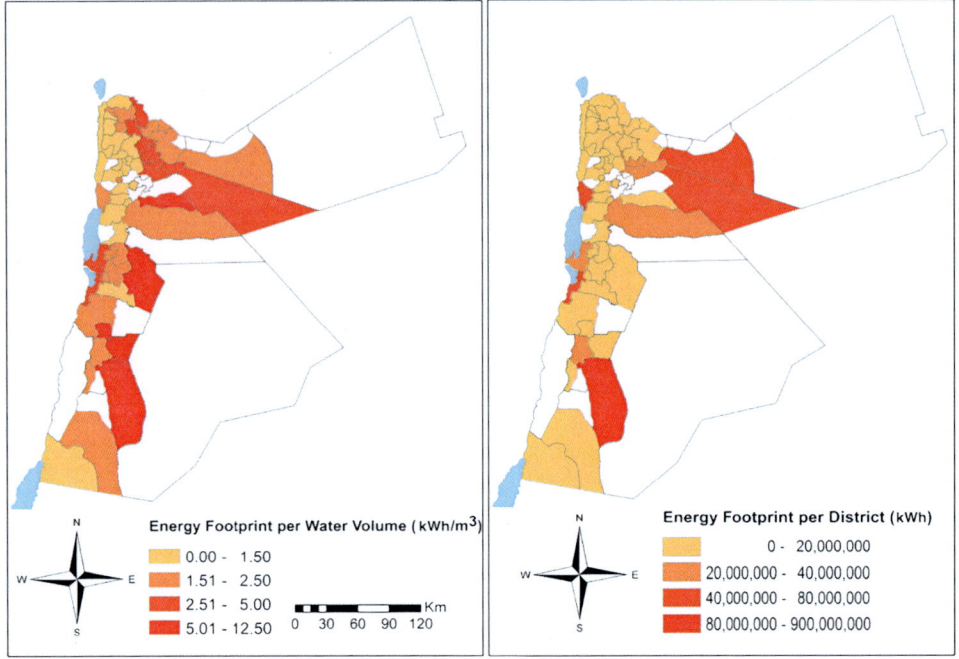

Figure 8. Energy footprint per cubic metre of water used per district and total annual energy footprint per district.

Conclusions and policy implications

In light of Jordan's usage of blue and green water, the concern is of course with the country's extremely limited water resources. The population was estimated at 6.5 million in 2013, with a population growth rate of 2.2 (Jordan Department of Statistic, 2013). It comes as no surprise that its per capita water supply for all uses in 2013 was less than 15% of what is recommended by international standards (Ministry of Water and Irrigation, 2013). Further stress is put upon the country's limited water resources by the large numbers of refugees from both Syria and Iraq. In its planning figures for January 2015, the UN's refugee agency estimates that 811,070 refugees and asylum-seekers (747,360 from Syria and 58,050 from Iraq) are living in Jordan, and these numbers are expected to increase (United Nations High Commissioner for Refugees, 2015). A majority of these refugees are in the north-western portion of Jordan, meaning that they are also drawing upon the country's much-diminished groundwater aquifers. With potential impacts from climate change also leading to less rainfall with each coming year, for Jordan to supply its people with even the minimal requirements of water will continue to be a great challenge.

It is additionally necessary to view the virtual water usage numbers in light of how much is being put towards produce that is exported from Jordan. To put this in perspective, the Disi Aquifer project currently supplies Amman with 100 MCM/y (Ministry of Water and Irrigation, 2012a; Salameh, Alraggad, & Tarawneh, 2014). This is a non-renewable blue water resource and it has to be pumped from 325 km south of Amman, with an altitude change of roughly 900 m, imposing a big energy footprint. The project is necessary because Jordan lacks other exploitable blue water resources, namely the limited surface water supply and the much-depleted groundwater resources being used for agriculture in the Desert region. As noted earlier, just over twice as much of Jordan's blue water is being exported (206 MCM) as is being exploited from the Disi Aquifer (100 MCM).

This raises the policy question of whether Jordan should be producing this quantity of fruits and vegetables, either for export or for its own use, with precious blue water resources that are needed in other sectors. While from a water resources point of view the answer might be no, there are practical reasons why Jordan continues along this path. Agriculture and its related activities and businesses employ roughly 29% of Jordan's labour force (Al-Karablieh, 2012). This number includes those involved in livestock production. Crop and livestock production is a sector that many Jordanians and immigrant labourers rely on for their livelihoods. There are also those in the market involved in the trading of these goods, and their livelihoods depend upon a robust Jordanian agricultural sector and the exporting of food commodities. So one must think of water, as Wichelns (2011) advises, as just one 'input' among many involved in the export and use of a good outside of Jordan. Additionally, as Alqadi and Kumar (2014) argue, Jordan faces a food security risk in the future due to its population growth and reliance on food imports that are constantly vulnerable to price increases, making within-country agricultural production a potentially vital strength.

It is also the case, as Yorke (2013) points out, that this trading in virtual water provides support for the continued functioning of the agricultural sector and those involved with it. Considering the agricultural sector's less productive circumstances as compared to other sectors (only 3.5% of the country's GDP), this support allows for the continuation of current perhaps inefficient uses of water and how these waters are governed. Barham (2012) suggests that the loyalty of powerful actors and groups in these sectors is desired by the state to the extent that water and land resources are used to gain support from diverse constituencies. While economic reforms should have occurred to shift energy (including water resources) and the workforce to more economically productive sectors, it was more politically feasible to maintain the status quo.

One way to approach Jordan's high usage of such limited blue water, especially as it contributes so much to exported goods and thus the exporting of Jordan's water resources in the form of virtual water, is to again return to the comparison between the blue water that is used within the Jordan Valley and that which is used in the Desert. Not only does the Jordan Valley use surface water instead of groundwater, but it is also located below sea level, allowing for a moderate climate in the winter months when it can take advantage of the local and global markets. More importantly with regard to energy usage, the low altitude of the Jordan Valley aids water flow into valley farms from the Yarmouk and Jordan Rivers, as well as from the side valleys, using largely the force of gravity. Pumping is used in certain parts of the water distribution network in the Jordan Valley, but far less than in the Desert, where pumping is necessary to lift and distribute groundwater. This energy usage is supported by the results found for the energy footprints by district. In addition, the wastewater that is used in the Jordan Valley is transported primarily via gravity from the treatment plant in Khirbat as-Samra to King Talal Dam and then to Jordan Valley farmlands. Along this system, treated wastewater (recycled blue water) is used to generate hydropower as it flows out of the King Talal Dam.

Considering that Jordan imports 97% of its energy from neighbouring countries and that these supplies are largely in the form of non-renewable fossil fuels and natural gas, fluctuating energy prices will always have a large impact on food production. This would explain why such a large percentage of Jordan's budget is spent on energy resources (Al-Ghandoor, 2013; Ministry of Energy and Mineral Resources, 2013). It is all the more important to conserve energy and find those water resources and areas that require less energy in order to make Jordan that much less vulnerable to international energy prices and the potential for social and political unrest. Also, it is worth considering that in those areas with the highest energy use (the Desert region), there is also a high solar energy potential. Meisen and Tatum (2011) report that on a Jordan-wide scale, the country receives an average of 2700 kwh of sunlight per square metre of land per year, so the potential is present in all regions.

At the social level, and getting back to a comparison between regions in Jordan, the Jordan Valley is the better option for sustainable agriculture because unlike in the Desert, where agriculture is more commercialized and unattached to its surrounding villages and environment, agriculture in the Jordan Valley is very much a part of the social fabric of neighbouring villages and people's livelihoods. Jordan Valley communities are closely connected to agricultural production in direct and indirect ways, with many working on farms as owners, engineers or labourers. Others work in related businesses that provide transportation services and agricultural inputs to farms. And still others are employees in the Jordan Valley Authority or the water user associations that aid in water distribution in some select areas.

Economically speaking, as Al-Karablieh (2012) has pointed out in his work on the water value of different crops and in different locations in Jordan, the Jordan Valley is where the most value can be reaped for the water used. He shows that winter vegetables, grown primarily in the Jordan Valley, have the highest water value, or the most revenue per cubic metre of water used. His study also shows that Jordan Valley farmers in general achieve the highest returns on water compared to other agricultural regions and that the returns on surface water are higher than those for groundwater.

Challenges do still exist in the Jordan Valley. While it maintains the more optimal environment for irrigated production compared with the Desert, problems of soil salinity occur as a consequence of high evaporation rates, exacerbated in some places by unwise use of treated wastewater and fertilizers or pesticides (Alqadi, Kumar, & Jarrah, 2013). The use of desalinated brackish water from wells, or its mixture with river water or treated wastewater, is also a problem for the same reason: increased soil salinity. In addition, the increased extraction of groundwater could result in the same energy-use problems present in the Desert. With regard to crop choice in the Jordan Valley, Norton and Jabarin (2006) have suggested focusing on those in each region with a higher value per cubic metre of water, such as potatoes in the North, greenhouse-grown cucumbers and tomatoes in the Middle, greenhouse-grown cucumbers and tomatoes plus squash in Karama, and watermelon and string beans in the South. Al-Karablieh and Jabarin (2012) add the much-needed caveat that the crops should also be those that require less water in general, such as cucumbers and green beans, as compared to the more water-intensive tomatoes. This is a tricky issue in the Jordan Valley, where many farmers do not desire or do not have the ability to change their cropping pattern even if alternative patterns could result in higher earnings and less water usage. In sum, it would be appropriate to consider more demanding best management practices in the use of treated wastewater (recycled blue water) and in the monitoring of groundwater use in the Jordan Valley, as well as better demand management strategies for farmers.

With regard to the Highlands and its use of green water, there is a potential for dryland cropping in some areas, thus harnessing more of Jordan's virtual green water potential. In those areas of the Desert that have useful levels of effective rainfall (green water), it would be optimal from a virtual water perspective if crop patterns could change to more drought-tolerant or rain-dependent species, thus taking advantage of this virtual green water instead of depleting Jordan's blue virtual water. One qualification to add is that some

(Al-Bakri, Suleiman, Abdulla, & Ayad, 2010) see climate change as leading to lower annual precipitation in the Highlands, making for a more vulnerable situation in the long run if there is greater dependence on rainfall for crops. These circumstances again highlight the need to develop more efficient water-use practices and adopt more drought-resilient and productive crops.

Acknowledgements

We thank Jordan's Ministry of Agriculture and Ministry of Water and Irrigation for the data that they provide and share, without which this study would not have been possible. The Ministry of Agriculture's detailed agricultural survey and the Ministry of Water and Irrigation's water budget and continued recording of water distribution numbers were immensely helpful to this project. We also express our gratitude for the work of the Deutsche Gesellschaft für Internationale Zusammenarbeit within Jordan's water sector. Their analyses have been invaluable for their attention to detail not just with regard to the technical aspects of water management but also with regard to the social, economic and environmental impacts of Jordan's actions within this domain.

Disclosure statement

No potential conflict of interest was reported by the authors.

References

Abdel Jabbar, S., & Sobh, A. (2012). *Detailed study of water requirements for fruit trees and vegetables grown in the Jordan Valley*. Amman: Deutsche Gesellschaft für Internationale Zusammenarbeit GmbH (GIZ).

Abdel Khaleq, R. A. (2011). Water soft path analysis – Jordan case. In U. Uhlig (Ed.), *Current issues of water management* (pp. 287–318). Intech. Retrieved from http://www.intechopen.com/books/current-issues-of-water-management/water-soft-path-analysis-jordan-case

Abu-Awwad, A. M., & Blair, S. (2013). Economic efficiency of water use by irrigated crops in Al'Azraq area. *Jordan Journal of Agricultural Sciences, 9*, 525–543. Retrieved from https://journals.ju.edu.jo/JJAS/article/viewFile/5461/3466

Abu-Sharar, T. M. (2006). The challenges of land and water resources degradation in Jordan: Diagnosis and solutions. In W. G. Kepner, J. L. Rubio, D. A. Mouat, & F. Pedrazzini (Eds.), *Desertification in the Mediterranean: A security issue* (pp. 201–226). The Netherlands: Springer.

Al-Bakri, J. T., Salahat, M., Suleiman, A., Suifan, M., Hamdan, M., Khresat, S., & Kandakji, T. (2013). Impact of climate and land use changes on water and food security in Jordan: Implications for transcending 'the tragedy of the commons'. *Sustainability, 5*, 724–748. doi:10.3390/su5020724

Al-Bakri, J. T., Suleiman, A., Abdulla, F., & Ayad, J. (2010). Potential impact of climate change on rainfed agriculture of a semi-arid basin in Jordan. *Physics and Chemistry of the Earth, Parts A/B/C, 36*, 125–134. Retrieved from http://www.journals.elsevier.com/physics-and-chemistry-of-the-earth/

Al-Eisawi, D. (2012). Conservation of natural ecosystems in Jordan. *Pakistan Journal of Botany, 44*, 95–99. Retrieved from http://www.pakbs.org/pjbot/PDFs/44%28SI2%29/13.pdf

Al-Ghandoor, A. (2013). Evaluation of energy use in Jordan using energy and exergy analyses. *Energy and Buildings, 59*, 1–10. Retrieved from http://www.sciencedirect.com/science/article/pii/S0378778812006846 10.1016/j.enbuild.2012.12.035

Al-Karablieh, E. (2012). *ISSP water valuation study: Disaggregated economic value of water in industry and irrigated agriculture in Jordan*. Amman: Institutional Support and Strengthening Program, United States Agency for International Development.

Al-Karablieh, E., & Jabarin, A. S. (2012). The water perspective in the Jordanian agricultural export competitiveness: "An analysis of selected products in Jordan Valley". Paper presented at 13 Scientific Day Conference, Amman, Jordan. Retrieved from http://sw15.rss.jo/files/%D9%83%

D8%B1%D8%A7%D8%A8%D9%84%D9%8A%D8%A9Water%20Perspective%20Paper% 2018-09-2010.doc

AlAyyash, S., Al-Adamat, R., & Al-Meshan, O. (2012). Application of Geo-informatics in mapping sites appropriate for the cultivation of forage in Jordan's Badia (Desert) region. *Surveying and Land Information Science*, *72*, 79–85. Retrieved from http://www.g-lis.org/education/ surveying-and-land-information-science-salis-journal/

Aldaya, M. M., & Llamas, M. R. (2008). *Water footprint analysis for the Guadiana river basin* (Value of Water Research Report Series No. 35). The Netherlands: UNESCO–IHE, Delft.

Aldaya, M. R. M., Martínez-Santos, P., & Llamas, M. (2010). Incorporating the water footprint and virtual water into policy: Reflections from the Mancha occidental region, Spain. *Water Resources Management*, *24*, 941–958. doi:10.1007/s11269-009-9480-8

Alqadi, K. A., & Kumar, L. (2014). Water policy in Jordan. *International Journal of Water Resources Development*, *30*, 322–334. doi:10.1080/07900627.2013.876234

Alqadi, K. A., Kumar, L., & Jarrah, A. (2013). Changing demographics, expanding urban areas and modified agricultural extents and their impacts on water availability and water quality in Jordan. *African Journal of Agricultural Research*, *8*, 3193–3201. doi:10.5897/AJAR12.875

Antonelli, M., Roson, R., & Sartori, M. (2012). Systemic input-output computation of green and blue virtual water 'flows' with an illustration for the Mediterranean region. *Water Resources Management*, *26*, 4133–4146. doi:10.1007/s11269-012-0135-9

Barham, N. (2012). *Is good water governance possible in a Rentier state? The case of Jordan* (Analysis). Odense: Center for Mellemøststudier, University of Southern Denmark. Retrieved from http://static.sdu.dk/mediafiles/F/4/0/%7BF40EA11F-48A8-4EF0-A0AE-2A25EC390AF 4%7DNB0512.pdf

Berger, M., & Finkbeiner, M. (2010). Water footprinting: How to address water use in life cycle assessment? *Sustainability*, *2*, 919–944. Retrieved from http://www.mdpi.com/2071-1050/2/4/ 91910.3390/su2040919

Chapagain, A. K., & Hoekstra, A. Y. (2003). *Virtual water flows between nations in relation to trade in livestock and livestock products* (Value of Water Research Report Series No. 13). The Netherlands: UNESCO-IHE, Delft.

Chapagain, A. K., & Hoekstra, A. Y. (2004). *Water footprints of nations, Vol. 1: Main report* (Value of Water Research Report Series No. 16) (Vol. 16). The Netherlands: UNESCO-IHE, Delft.

Chapagain, A. K., & Hoekstra, A. Y. (2004). *Water footprints of nations, Vol. 2: Appendices* (Value of Water Research Report Series No. 16) (Vol. 16). The Netherlands: UNESCO-IHE, Delft.

Daniels, P. L., Lenzen, M., & Kenway, S. J. (2011). The ins and outs of water use – a review of multi-region input-output analysis and water footprints for regional sustainability analysis and policy. *Economic Systems Research*, *23*, 353–370. doi:10.1080/09535314.2011.633500

Dastane, N. G. (1978). Effective rainfall. Irrigation and Drainage Paper 25. Rome: Food and Agriculture Organization. Retrieved from http://www.fao.org/docrep/x5560e/x5560e00. htm#Contents

Demilecamps, C., & Sartawi, W. (2010). *Farming in the desert: Analysis of the agricultural situation in the Azraq basin*. Amman: Deutsche Gesellschaft für Internationale Zusammenarbeit GmbH (GIZ).

Department of Statistics. (2013). Jordan in figures 2013. Retrieved from http://www.dos.gov.jo/dos_ home_e/main/linked-html/jordan_no.htm

Dietzenbacher, E., & Velázquez, E. (2007). Analysing Andalusian virtual water trade in an input– output framework. *Regional Studies*, *41*, 185–196. doi:10.1080/00343400600929077

Fader, M., Rost, S., Müller, C., Bondeau, A., & Gerten, D. (2010). Virtual water content of temperate cereals and maize: Present and potential future patterns. *Journal of Hydrology*, *384*, 218–231. Retrieved from http://www.journals.elsevier.com/journal-of-hydrology/10.1016/j.jhydrol.2009. 12.011

Freiwan, M., & Kadioglu, M. (2008). Spatial and temporal analysis of climatological data in Jordan. *International Journal of Climatology*, *28*, 521–535. doi:10.1002/joc.1562

Guan, D., & Hubacek, K. (2007). Assessment of regional trade and virtual water flows in China. *Ecological Economics*, *61*, 159–170. doi:10.1016/j.ecolecon.2006.02.022

Hellegers, P., Zilberman, D., Steduto, P., & McCornick, P. (2008). Interactions between water, energy, food and environment: Evolving perspectives and policy issues. *Water Policy*, *10*(S1), 1–10. doi:10.2166/wp.2008.048

Hoekstra, A. Y., & Chapagain, A. K. (2008). *Globalization of water: Sharing the planet's freshwater resources*. Oxford: Blackwell.

Hoekstra, A. Y., & Hung, P. Q. (2002). *Virtual water trade: A quantification of virtual water flows between nations in relation to international crop trade* (Value of Water Research Report Series 12). The Netherlands: UNESCO-IHE, Delft.

International Resources Group. (2013, updated 2014). National strategic wastewater master plan – final report. Institutional Support & Strengthening Plan, United States Agency for International Development, Amman. Retrieved from http://isspjordan.org/files/upload/resources/887b1bd5c49d87e23fbf18201db3326c.pdf

Madanat, H. J. (2010). Land tenure in Jordan. *Land Tenure Journal, 1*, 143–170. Retrieved from http://www.fao.org.webtranslate-widget.systransoft.com/nr/tenure/land-tenure-journal/index.php/LTJ/article/view/12/6

McCornick, P. G., Awulachew, S. B., & Abebe, M. (2008). Water-food-energy-environment synergies and tradeoffs: Major issues and case studies. *Water Policy, 10*, 23–36. doi:10.2166/wp.2008.050

Meisen, P., & Tatum, J. (2011). *The water-energy nexus in the Jordan river basin: The potential for building peace through sustainability*. San Diego, CA: Global Energy Network Institute. Retrieved from http://www.geni.org/globalenergy/research/water-energy-nexus-in-the-jordan-river-basin/the-jordan-river-basin-final-report.pdf

Mercy Corps. (2014). *Tapped out: Water scarcity and refugee pressures in Jordan*. Amman: Mercy Corps.

Meteorological Department. (2012). *General recommendations (report)*. Amman: Jordan Meteorological Department.

Ministry of Agriculture. (2012). *Annual statistics report 2012*. Amman: Department of Information, Directorate of Information Technology, Ministry of Agriculture. Retrieved from http://www.moa.gov.jo/ar-jo/home.aspx

Ministry of Energy and Mineral Resources. (2012). *Annual report 2012*. Amman: Ministry of Energy and Mineral Resources. Retrieved from http://www.memr.gov.jo/LinkClick.aspx?fileticket=4b99mUCxv1M%3d&tabid=111

Ministry of Energy and Mineral Resources. (2013). *Annual report 2013*. Amman: Ministry of Energy and Mineral Resources. Retrieved from http://www.memr.gov.jo/LinkClick.aspx?fileticket=26IdFN-aosQ%3d&tabid=111

Ministry of Water and Irrigation. (2009). *Water for life: Jordan's water strategy 2008–2022*. Amman: Ministry of Water and Irrigation.

Ministry of Water and Irrigation. (2012a). *Annual report 2012*. Amman: Ministry of Water and Irrigation. Retrieved from http://www.mwi.gov.jo/sites/en-us/Annual%20Reports/MWI%202012%20English%20Report.pdf

Ministry of Water and Irrigation. (2012b). *Jordanian water budget 2011–2012*. Amman: Ministry of Water and Irrigation.

Ministry of Water and Irrigation. (2013). *Jordan water sector facts and figures 2013*. Amman: Ministry of Water and Irrigation. Retrieved from http://www.mwi.gov.jo/sites/en-us/Documents/W.%20in%20Fig.E%20FINAL%20E.pdf

Molden, D., Oweis, T. Y., Pasquale, S., Kijne, J. W., Hanjra, M. A., Bindraban, P. S., & … Zwart, S. (2007). Pathways for increasing agricultural water productivity. In *Water for food, water for life. A comprehensive assessment of water management in agriculture* (pp. 279–310). London, Colombo: International Water Management Institute.

Montesinos, P., Camacho, E., Campos, B., & Rodríguez-Díaz, J. A. (2011). Analysis of virtual irrigation water. Application to water resources management in a Mediterranean river basin. *Water Resources Management, 25*, 1635–1651. doi:10.1007/s11269-010-9765-y

Mourad, K. A., Gaese, H., & Jabarin, A. S. (2010). Economic value of tree fruit production in Jordan Valley from a virtual water perspective. *Water Resources Management, 24*, 2021–2034. Retrieved from http://link.springer.com/journal/1126910.1007/s11269-009-9536-9

National Electric Power Company. (2013). *Annual report 2013*. Amman: National Electric Power Company. Retrieved from http://www.nepco.com.jo/store/docs/web/2013_en.pdf

Norton, R. D., & Jabarin, A. S. (2006). *Toward more efficient agricultural production and marketing in Jordan*. Amman: United States Agency for International Development. Retrieved from http://www.ncare.gov.jo/OurNCAREPages/PROJECTMENU/RelatedPages/KAFAA/Kafa%27a%20assessment/CR-12.%20Toward%20more%20efficient%20marketing.pdf

Parker, C. (2012). *Urbanization and Its Effect on Rural Jordan: An Analysis of Badia to Amman Resource Flow*. (Unpublished SIT Study Abroad thesis). San Francisco State University, San Francisco, California.

Roth, D., & Warner, J. (2008). Virtual water: Virtuous impact? The unsteady state of virtual water. *Agriculture and Human Values*, *25*, 257–270. doi:10.1007/s10460-007-9096-7

Salameh, E. (2008). Over-exploitation of groundwater resources and their environmental and socio-economic implications: The case of Jordan. *Water International*, *33*, 55–68. doi:10.1080/02508060801927663

Salameh, E., Alraggad, M., & Tarawneh, A. (2014). Disi water use for irrigation – a false decision and its consequences. *CLEAN – Soil, Air, Water*, *42*, 1681–1686. Retrieved from http://onlinelibrary.wiley.com/journal/10.1002/%28ISSN%291863-0669

Siebert, S., & Döll, P. (2010). Quantifying blue and green virtual water contents in global crop production as well as potential production losses without irrigation. *Journal of Hydrology*, *384*, 198–217. Retrieved from http://www.journals.elsevier.com/journal-of-hydrology/10.1016/j.jhydrol.2009.07.031

United Nations High Commissioner for Refugees. (2015). 2015 UNHCR country operations profile – Jordan. Retrieved from http://www.unhcr.org/pages/49e486566.html

Venot, J. P., & Molle, F. (2008). Groundwater depletion in the Jordan highlands: can pricing policies regulate irrigation water use? *Water Resources Management*, *22*, 1925–1941. doi:10.1007/s11269-008-9260-x

Venot, J. P., Molle, F., & Hassan, Y. (2007). *Irrigated agriculture, water pricing and water savings in the Lower Jordan River Basin (in Jordan)* (Comprehensive Assessment of Water Management in Agriculture Research Report 18). Colombo: International Water Management Institute.

Wichelns, D. (2001). The role of 'virtual water' in efforts to achieve food security and other national goals, with an example from Egypt. *Agricultural Water Management*, *49*, 131–151. Retrieved from http://www.journals.elsevier.com/agricultural-water-management/10.1016/S0378-3774(00)00134-7

Wichelns, D. (2010). Virtual water: A helpful perspective, but not a sufficient policy criterion. *Water Resources Management*, *24*, 2203–2219. doi:10.1007/s11269-009-9547-6

Wichelns, D. (2011). Do the virtual water and water footprint perspectives enhance policy discussions? *International Journal of Water Resources Development*, *27*, 633–645. doi:10.1080/07900627.2011.619894

Yang, H., Wang, L., Abbaspour, K. C., & Zehnder, A. J. B. (2006). Virtual water highway: Water use efficiency in global food trade. *Hydrology and Earth System Sciences Discussions*, *3*(1), 1–26. http://www.hydrol-earth-syst-sci-discuss.net/papers_in_open_discussion.html 10.5194/hessd-3-1-2006

Yorke, V. (2013). *Politics matter: Jordan's Path to Water Security Lies Through Political Reforms and Regional Cooperation* (Working Paper No. 2013/19). Bern: National Centre of Competence in Research on Trade Regulation, World Trade Institute of the University of Bern, Switzerland.

Zhang, Z. Y., Yang, H., Shi, M. J., Zehnder, A. J. B., & Abbaspour, K. C. (2011). Analyses of impacts of China's international trade on its water resources and uses. *Hydrology and Earth System Sciences*, *15*, 2871–2880. doi:10.5194/hess-15-2871-2011

Index